BASIC RESEARCH OPPORTUNITIES IN Earth Science

Committee on Basic Research Opportunities in the Earth Sciences

Board on Earth Sciences and Resources

Commission on Geosciences, Environment, and Resources

National Research Council

NATIONAL ACADEMY PRESS
Washington, D.C.

NOTICE: The project that is the subject of this report was approved by the Governing Board of the National Research Council, whose members are drawn from the councils of the National Academy of Sciences, the National Academy of Engineering, and the Institute of Medicine. The members of the committee responsible for the report were chosen for their special competences and with regard for appropriate balance.

This study was supported by Grant No. EAR-9809585 between the National Academy of Sciences and the National Science Foundation. Any opinions, findings, conclusions, or recommendations expressed in this publication are those of the author(s) and do not necessarily reflect the view of the National Science Foundation.

International Standard Book Number 0-309-07133-X
Library of Congress Card Number 00-111596

Additional copies of this report are available from:

National Academy Press
2101 Constitution Ave., NW
Box 285
Washington, DC 20055
800-624-6242
202-334-3313 (in the Washington Metropolitan Area)
http://www.nap.edu

Cover: Spatial scales relevant to Earth science processes. *Lower middle*: scanning electron photomicrograph of *Streptomyces sp.* growing on a polished hornblende surface, showing 400 to 600 nm-wide hyphae. SOURCE: H. Buss, Pennsylvania State University. *Bottom*: mid-crustal exposure of an early Paleozoic subduction zone in northwestern Norway. SOURCE: L. Royden, Massachusetts Institute of Technology.*Top left*: synthetic aperture radar image showing the postseismic displacement (10 mm interval) of the 1999 Hector Mine earthquake, California. SOURCE: data from the European Space Agency Satellite ERS-2 were acquired and processed by D. Sandwell, L. Sichoix, A. Jacobs, R. Scharroo, B. Minster, Y. Bock, P. Jameson, E. Price, and H. Zebker, Scripps Institution of Oceanography. *Upper middle*: images of mountains and calderas on Jupiter's volcanic moon Io taken by NASA's Galileo spacecraft. SOURCE: Jet Propulsion Laboratory.

Printed in the United States of America

THE NATIONAL ACADEMIES

National Academy of Sciences
National Academy of Engineering
Institute of Medicine
National Research Council

The **National Academy of Sciences** is a private, nonprofit, self-perpetuating society of distinguished scholars engaged in scientific and engineering research, dedicated to the furtherance of science and technology and to their use for the general welfare. Upon the authority of the charter granted to it by the Congress in 1863, the Academy has a mandate that requires it to advise the federal government on scientific and technical matters. Dr. Bruce M. Alberts is president of the National Academy of Sciences.

The **National Academy of Engineering** was established in 1964, under the charter of the National Academy of Sciences, as a parallel organization of outstanding engineers. It is autonomous in its administration and in the selection of its members, sharing with the National Academy of Sciences the responsibility for advising the federal government. The National Academy of Engineering also sponsors engineering programs aimed at meeting national needs, encourages education and research, and recognizes the superior achievements of engineers. Dr. William A. Wulf is president of the National Academy of Engineering.

The **Institute of Medicine** was established in 1970 by the National Academy of Sciences to secure the services of eminent members of appropriate professions in the examination of policy matters pertaining to the health of the public. The Institute acts under the responsibility given to the National Academy of Sciences by its congressional charter to be an adviser to the federal government and, upon its own initiative, to identify issues of medical care, research, and education. Dr. Kenneth I. Shine is president of the Institute of Medicine.

The **National Research Council** was organized by the National Academy of Sciences in 1916 to associate the broad community of science and technology with the Academy's purposes of furthering knowledge and advising the federal government. Functioning in accordance with general policies determined by the Academy, the Council has become the principal operating agency of both the National Academy of Sciences and the National Academy of Engineering in providing services to the government, the public, and the scientific and engineering communities. The Council is administered jointly by both Academies and the Institute of Medicine. Dr. Bruce M. Alberts and Dr. William A. Wulf are chairman and vice chairman, respectively, of the National Research Council.

COMMITTEE ON BASIC RESEARCH OPPORTUNITIES IN THE EARTH SCIENCES

THOMAS H. JORDAN, *Chair*, University of Southern California, Los Angeles
GAIL M. ASHLEY, Rutgers University, Piscataway, New Jersey
MARK D. BARTON, University of Arizona, Tucson
STEPHEN J. BURGES, University of Washington, Seattle
KENNETH A. FARLEY, California Institute of Technology, Pasadena
KATHERINE H. FREEMAN, The Pennsylvania State University, University Park
RAYMOND JEANLOZ, University of California, Berkeley
CHARLES R. MARSHALL, Harvard University, Cambridge, Massachusetts
JOHN A. ORCUTT, Scripps Institution of Oceanography, La Jolla, California
FRANK M. RICHTER, University of Chicago, Illinois
LEIGH H. ROYDEN, Massachusetts Institute of Technology, Cambridge
CHRISTOPHER H. SCHOLZ, Lamont-Doherty Earth Observatory, Palisades, New York
NOEL TYLER, The University of Texas, Austin
LAWRENCE P. WILDING, Texas A&M University, College Station

National Research Council Staff

ANNE M. LINN, Senior Staff Officer
VERNA J. BOWEN, Administrative Assistant

BOARD ON EARTH SCIENCES AND RESOURCES

COMMISSION ON GEOSCIENCES, ENVIRONMENT, AND RESOURCES

Acknowledgments

This report has been reviewed in draft form by individuals chosen for their diverse perspectives and technical expertise, in accordance with procedures approved by the NRC's Report Review Committee. The purpose of this independent review is to provide candid and critical comments that will assist the institution in making its published report as sound as possible and to ensure that the report meets institutional standards for objectivity, evidence, and responsiveness to the study charge. The review comments and draft manuscript remain confidential to protect the integrity of the deliberative process. We wish to thank the following individuals for their review of this report:

Albert Bally, Department of Geology and Geophysics, Rice University
Vitelmo V. Bertero, Earthquake Engineering Research Center, University of California, Berkeley
Jeremy Bloxham, Department of Earth and Planetary Sciences, Harvard University
David L. Donoho, Department of Statistics, Stanford University
Thomas Dunne, Donald Bren School of Environmental Science and Management, University of California, Santa Barbara
Wilford R.Gardner, College of Natural Resources, University of California, Berkeley
Paul L. Koch, Department of Earth Sciences, University of California, Santa Cruz
George McGill, Department of Geosciences, University of Massachusetts
Peter Molnar, Falmouth, Massachusetts
Karl Turekian, Kline Geology Laboratory, Yale University

Although the reviewers listed above have provided many constructive comments and suggestions, they were not asked to endorse the conclusions or recommendations nor did they see the final draft of the report before its release. The review of this report was overseen by Mary Lou Zoback, appointed by the Commission on Geosciences, Environment, and Resources and Steven M. Stanley, appointed by the NRC's Report Review Committee, who were responsible for making certain that an independent examination of this report was carried out in accordance with institutional procedures and that all review comments were carefully considered. Responsibility for the final content of this report rests entirely with the authoring committee and the institution.

Preface

This report summarizes the findings and recommendations of the Committee on Basic Research Opportunities in the Earth Sciences. The committee was charged by the National Science Foundation (NSF) to undertake the following tasks:

- identify high-priority research opportunities in the Earth sciences, emphasizing the connections between traditional solid-Earth science disciplines such as geodynamics, geology, and geochemistry and other disciplines such as hydrology, biology, and oceanography;
- discuss research opportunities of interest to other government agencies, industry, and international partners, to the extent that they are germane to the responsibilities of NSF's Earth Science Division (EAR); and
- explore linkages between research and societal needs.

In keeping with its charge, the committee did not review the existing EAR program or other federal research programs. Rather, this report focuses on new research areas that could be added to the EAR solid-Earth science and hydrology portfolio. Similarly, because EAR funds are limited, the committee did not emphasize research directions that are funded predominantly by other NSF divisions, such as paleoceanography and marine geophysics (Ocean Sciences Division) and paleoclimatology (Atmospheric Sciences Division and Office of Polar Programs).

Previous National Research Council (NRC) reports have significantly helped to shape NSF activities. Prior to 1983, EAR directed all of its funds to individual investigators through core research programs. On the recommendations of *Opportunities for Research in the Geological Sciences*[1] and *Research Briefings*,[2] EAR created a variety of cross-disciplinary programs, including Instrumentation and Facilities and Continental Dynamics.

The National Research Council published its last major assessment of Earth science in 1993. *Solid-Earth Science and Society*[3] documented progress in Earth science, its technology drivers, the status of its constituent disciplines, a host of significant unsolved problems, and many outstanding research opportunities. It also articulated the fundamental importance of Earth science in a globalized, high-technology society. Much of what it said seven years ago remains fresh and applicable today.

In conducting this study, the committee found it necessary to survey a wide range of topics across a broad spectrum of disciplines, which required input from many individuals and groups. Before its first meeting, the committee sponsored symposia at the annual Geological Society of America and American Geophysical Union meetings. Presenters were asked to provide a 10-year vision of the research opportunities in their field. In addition, the committee requested the following information from department heads at universities and colleges, professional societies, and federal agencies with a significant Earth science component:

- the 10-year outlook for the Earth sciences, including possible linkages with other disciplines;
- the scale of activities suitable for conducting Earth science, including the roles of individual investigators, major facilities, and "system-level" research; and
- the programmatic mechanisms and level of funding needed from the NSF and other agencies.

Federal agencies with major Earth science programs—the National Science Foundation, U.S. Geological Survey, Department of Energy, and National Aeronautics and Space Administration—also provided programmatic information and "lessons learned" from past collaborations. Finally, the committee reviewed a variety of workshop reports and white papers, which were sponsored by NSF and/or professional societies in the past two years. The titles of workshop reports and symposia abstracts and the names of

---

[1]NRC, *Opportunities for Research in the Geological Sciences*. National Academy Press, Washington, D.C., 95 pp., 1983.

[2]NRC, *Research Briefings 1983*. National Academy Press, Washington, D.C., 99 pp., 1983.

[3]NRC, *Solid-Earth Sciences and Society*. National Academy Press, Washington, D.C., 346 pp., 1993.

survey respondents are listed in Appendix B. Many of the conclusions and recommendations reached by the committee reflect the ideas articulated in these thoughtful contributions by numerous members of the geoscience community.

The committee also acknowledges the following individuals, who briefed the committee, provided detailed programmatic information, or contributed in other ways to the committee process: Morris Aizenman, Jill Banfield, Steven Bohlen, Joe Burns, Robert Corell, Bill Dietrich, Adam Dziewonski, John Grant, Ron Greeley, Richard Greenfield, Douglas James, Russel Kelz, Susan Kidwell, Ian MacGregor, Michael Mayhew, Michael Meyer, Mike Purdy, Garrison Sposito, Dave Stevenson, Dorothy Stout, Bruce Uminger, Daniel Weill, Clark Wilson, Nick Woodward, and Herman Zimmerman. Finally, the committee expresses its gratitude to the NRC study director, Anne Linn, for her considerable efforts in bringing the committee together and editing its report.

Thomas H. Jordan
Chair

Contents

Executive Summary

Earth science is a quest for fundamental knowledge about the origin, evolution, and future of the natural world. Opportunities in this science have been opened up by major improvements in techniques for reading the geological record of terrestrial change, capabilities for observing active processes in the present-day Earth, and computational technologies for realistic simulations of dynamic geosystems. The agenda for the next decade of basic research is to explore the planet—decipher its history, understand its current behavior, and predict its future—by exploiting and extending these capabilities. This research will contribute to five national imperatives: (1) discovery, use, and conservation of natural resources; (2) characterization and mitigation of natural hazards; (3) geotechnical support of commercial and infrastructure development; (4) stewardship of the environment; and (5) terrestrial surveillance for global security and national defense. Progress on these practical issues depends on basic research across the full spectrum of Earth science. The National Science Foundation (NSF), through its Earth Science Division (EAR), is the only federal agency that maintains significant funding for basic research in all the core disciplines of Earth science. The health of the EAR program is therefore central to a strong national effort in Earth science.

OPPORTUNITIES FOR BASIC RESEARCH

Basic research in Earth science encompasses a wide range of physical, chemical, and biological processes that interact and combine in complex ways to produce a hierarchy of terrestrial systems. EAR is currently sponsoring investigations on geosystems that range in geographic scale from global—climate, plate tectonics, and the core dynamo—to regional and local—

mountain belts and sedimentary basins, active fault networks, volcanoes, groundwater reservoirs, and soil systems. Research at all of these scales has been accelerated by a combination of conceptual advances and across-the-board improvements in observational capabilities and information technologies. The committee has identified six specific areas, organized here by proximity and scale, in which the opportunities for basic research are especially compelling:

1. *Integrative studies of the "Critical Zone,"* the heterogeneous, near-surface environment in which complex interactions involving rock, soil, water, air, and living organisms regulate the natural habitat and determine the availability of life-sustaining resources. Many science disciplines—hydrology, geomorphology, biology, ecology, soil science, sedimentology, materials research, and geochemistry—are bringing novel research tools to bear on the study of the Critical Zone as an integrated system of interacting components and processes. During the next decade, basic research will be able to address a wide spectrum of interconnected problems that bear directly on societal interests:

- terrestrial carbon cycle and its relationship to global climate change, including the temporal and spatial variability of carbon sources and sinks and the influence of weathering reactions,
- quantification of microbial interactions in mineral weathering, soil formation, the accumulation of natural resources, and the mobilization of nutrients and toxins,
- dynamics of the land-ocean interface, which governs how coastal ocean processes such as tides, waves, and currents interact with river drainage, groundwater flow, and sediment flux,
- coupling of the tectonic and atmospheric processes through volcanism, precipitation, fluvial processes, glacier development, and erosion, which regulate surface topography and influence climate on geological time scales, and
- formation of a geological record that encodes a four-billion-year history of Critical-Zone processes, including environmental variations caused by major volcanic episodes, meteorite impacts, and other extreme events.

2. *Geobiology*, the study of how life interacts with the Earth and how it has changed through geological time. By combining the powerful tools of genomics, proteinomics, and developmental biology with new techniques from geochemistry, mineralogy, stratigraphy, and paleontology, geobiologists are now better equipped to investigate a variety of fundamental problems:

- prebiotic molecules, origin of life, and early evolution,
- biological and environmental controls on species diversity, including ecological and biogeographic selectivity, causes of extinction and survival, and the nature of evolutionary innovation,
- response of organisms, communities, and ecosystems to environmental perturbations, including the role of extreme events in reshaping ecosystems and climate,
- biogeochemical interactions and cycling among organisms, ecosystems, and the environment, with applications to monitoring and remediating environmental degradation, and
- effects of natural and anthropogenic environmental change on the habitability of the Earth.

3. *Research on Earth and planetary materials*, which uses advanced instrumentation and theory to determine properties at the molecular level for understanding materials and processes at all scales relevant to planets. This field is being stimulated by enhanced research capabilities, such as synchrotron-beamlines for micro-diffraction and spectroscopy, experimental apparatus for accessing ultra-high pressures and temperatures, resonance techniques for precise measurements of elastic properties, quantum-mechanical simulations of complex minerals, and novel approaches to geomicrobiology and biomineralogy. A number of opportunities for basic research can be identified:

- biomineralization—natural growth of minerals within organisms, with applications to the development of synthetic analogs,
- characterization of extraterrestrial samples from Mars, comets, and interplanetary space,
- super-high pressure (terapascal) research, with applications to planetary and stellar interiors,
- nonlinear interactions and interfacial phenomena in rocks—strain localization, nonlinear wave propagation, fluid-mineral reactions, and coupling of chemical reactions to fracturing,
- nanophases and interfaces, including microbiology at interfaces and applications to the physics and chemistry of soils,
- quantum and molecular theory applied to minerals and their interfaces, and
- studies of granular media, including the nonlinear physics of soils and loose aggregates.

4. *Investigations of the continents*. New space-based geodetic techniques—the Global Positioning System and interferometric synthetic aperture

radar (InSAR)—are capable of mapping crustal deformation with centimeter-level precision, paving the way for advances in earthquake mechanics, volcano physics, and crustal rheology. Seismic tomography can now image the subsurface with enough horizontal resolution to observe how individual surface features are expressed at depth. These remote-sensing techniques, in combination with field mapping, deep continental drilling for in situ sampling and experimentation, and advanced laboratory analysis of rocks brought up from great depths, offer major opportunities to address basic questions regarding the three-dimensional structure and composition of the continents, the geologic record of continental formation and assembly, and the physical processes in continental deformation zones. Targets of this research include:

- mechanisms of active deformation, earthquake physics, coupling between brittle and ductile deformations, and fault-system dynamics and evolution,
- role of fluids in chemical, thermal, magmatic and mechanical processes, deep circulation systems in hydrothermal areas and sedimentary basins, and fluxes from the mantle,
- nature of the lower continental crust, its average composition and fluid content, processes of formation and development, and role as a mechanical decoupling layer, and
- deep structure of the continental lithosphere, its coupling to the underlying mantle, and implications for Earth evolution.

5. *Studies of the Earth's deep interior*, to define its structure, composition, and state, and to understand the machinery of mantle convection and the core dynamo. The quality and quantity of data are expanding at an extraordinary rate in many related fields—seismology, geomagnetic studies, geochemistry, and high-pressure research. Increased computational speeds and high-bandwidth networks have greatly facilitated the processing of very large data sets and the realistic modeling of deep-interior dynamics. Laboratory studies conducted at mantle and core conditions are now able to provide constraints on the physical and chemical conditions essential for the interpretation of numerical simulations. There are four primary areas of investigation:

- complex time-dependent flow patterns of solid-state mantle convection, which can be inferred by reconciling seismic tomographic and geochemical data using high-resolution numerical simulations,
- operation and interaction of mantle convection and the core dynamo over Earth history, which can be studied through multidisciplinary investigations of the core-mantle boundary,

- generation of the geomagnetic field, which can be investigated through realistic numerical simulations of the core dynamo, combined with recently available satellite and paleomagnetic data, and
- origin and evolution of the inner core and its role in the core dynamo, as revealed by the strong seismic heterogeneity and anisotropy discovered in the past few years.

6. *Planetary science*, which uses extraterrestrial materials, as well as astronomical, space-based, and laboratory observations, to investigate the origin, evolution, and present structure of planetary bodies, including the Earth. Telescopic observations of primitive objects in the solar system and of the planets orbiting distant stars are beginning to furnish unique data regarding the origin and evolution of the solar system. Current and planned space missions will provide unprecedented detail and coverage of the geology, topography, structure, and composition of many solar-system bodies. Within a decade, the first samples collected from Mars, a comet, an asteroid, and the Sun (via solar wind particles) will be returned to Earth for direct investigation. A proper interpretation of these data will require the application of Earth-science techniques and appropriate terrestrial comparisons. Such comparisons promise improved understanding of the Earth and solar system as a whole:

- Other planets furnish new environments for investigating the basic geological and geophysical processes operating on and within the Earth.
- Most planets preserve physical and chemical records of the early solar system that contains data on planetary evolution that no longer exists on Earth.
- Distinctive chemical and isotopic signatures from extraterrestrial samples are critical for furthering the understanding of the mixing, accretion, and differentiation of meteorite parent bodies and planets, including the Earth.

PRINCIPAL FINDINGS AND RECOMMENDATIONS

EAR has done an excellent job in maintaining the balance among core programs supporting investigator-driven disciplinary research, problem-focused programs of multidisciplinary research, and equipment-oriented programs for new instrumentation and facilities. The committee offers recommendations that address the evolving science requirements in all three of these programmatic areas. These recommendations pertain primarily to new mechanisms that will allow EAR to exploit research opportunities identified by the committee.

Long-Term Support of Investigator-Driven Science

EAR funding of research projects initiated and conducted by individual investigators and small groups of investigators is the single most important mechanism for maintaining and enhancing disciplinary strength in Earth science. Major investments are now justified in two promising fields. EAR should seek new funds for the long-term support of:

1. geobiology, to permit studies of the interactions between biological and geological processes, the evolution of life on Earth, and the geologic factors that have shaped the biosphere, and
2. investigator-initiated research on Earth and planetary materials to take advantage of major new facilities, advanced instrumentation and theory in an atomistic approach to properties and processes.

Outstanding research opportunities related to the study of the Critical Zone also warrant additional resources for established programs in hydrology and geology. The committee offers two primary recommendations:

- Owing to the significant opportunities for progress in the understanding of hydrologic systems, particularly through coordinated studies of the Critical Zone, EAR should continue to build programs in the hydrologic sciences.
- EAR should enhance multidisciplinary studies of the Critical Zone, placing special attention on strengthening soil science and the study of coastal zone processes.

To coordinate support for multidisciplinary studies, EAR should take the lead within NSF in devising a long-term strategy for funding research on the Critical Zone.

Mechanisms for Multidisciplinary Research

Understanding the behavior and evolution of complex terrestrial systems requires cooperative efforts in data collection as well as integrative studies to pull together diverse data sets and construct explanatory models. EAR has a very good record of sponsoring multidisciplinary research through its long-term core programs, particularly the Continental Dynamics Program, and a number of fixed-term special emphasis areas. The committee has identified several opportunities for strengthening the multidisciplinary aspects of Earth science.

EarthScope. This major NSF initiative, already in the advanced planning stage, will deploy four new observational systems: (1) *USArray*, for high-resolution seismological imaging of the structure of the crust and mantle beneath North America; (2) *San Andreas Fault Observatory at Depth*, for probing and monitoring the San Andreas Fault by deep drilling into the fault zone; (3) *Plate Boundary Observatory*, for measuring deformations of the western United States using strainmeters and ultraprecise geodesy; and (4) *InSAR*, for using satellite-based interferometric synthetic aperture radar to map surface deformations. EarthScope will contribute substantially to understanding the active tectonics and evolution of the continents, earthquake and volcanic hazards, and basic geodynamic processes operating in the Earth's deep interior. The scientific vision and goals of EarthScope are well articulated and have been developed with a high degree of community involvement.

- The committee strongly endorses the four observational components of the EarthScope initiative.

Existing programmatic elements within EAR furnish the mechanisms to support the basic science required for a successful EarthScope initiative, but only if funding is adequately augmented for basic disciplinary and multidisciplinary research.

Natural Laboratories. Demands are rising for EAR investments in natural laboratories, where terrestrial processes and systems can be studied through detailed field observations and in situ measurements in specially designated areas. This type of cooperative research is particularly suitable for studies of the Critical Zone, in which techniques from several disciplines must be coordinated to collect data sets that are spatially dense and temporally extended.

- EAR should establish an Earth Science Natural Laboratory (ESNL) Program, open to all problem areas and disciplines, with the objective of supporting long-term, multidisciplinary research at a number of promising sites within the United States and its territories.

Special Areas of Multidisciplinary Research. In addition to major facility-oriented initiatives, the committee suggests that EAR initiate fixed-term programs in two research areas—microorganisms in the environment and planetary science—that offer particular promise for significantly advancing scientific understanding through multidisciplinary studies:

- EAR should seek new resources to promote integrative studies of the way in which microorganisms interact with the Earth's surface environment, including present and past relationships between geological processes and the evolution and ecology of microbial life.

- To promote increased interactions between the Earth and planetary science research communities and to exploit the basic research opportunities arising in the study of solar and extrasolar planets, EAR should initiate a cooperative effort with the National Aeronautics and Space Administration (NASA) and NSF-Astronomy in planetary science.

Instrumentation and Facilities

The EAR Instrumentation and Facilities (I&F) Program has been highly successful, but it is under increasing stress from the rising costs of purchasing, operating, and maintaining state-of-the-art research equipment. To take advantage of novel technologies, EAR will have to expand the resources devoted to major facilities and observatories, as well as to individual laboratories. Technologies targeted for future investments might include neutron-scattering facilities, smart synchrotron beamlines, laser-based materials analysis, geochemical and geochronometric instrumentation, and mobile instrumentation for ground-based remote-sensing and biogeochemical analyses.

- EAR should seek more resources to support the growing need for new instrumentation, multiuser analytical facilities, and long-term observatories, and for ongoing support of existing equipment.
- The I&F program should encourage its user communities to identify research priorities and develop a consensus regarding how many laboratories are needed and how their operational costs should be apportioned among the EAR core programs, the I&F program, and participating academic institutions.

Education

To maintain its vitality, Earth science must attract talented new practitioners. The educational requirements for these practitioners are becoming more demanding, especially given the need to keep pace with the cross-disciplinary aspects of Earth science. Within EAR, there are many opportunities for blending education with basic research.

- EAR should institute training grants and expand its fellowship program to facilitate broad-based education for undergraduate and graduate students in the Earth sciences.

- EAR should establish postdoctoral and sabbatical-leave training programs to facilitate development of the cross-disciplinary expertise needed to exploit research opportunities in geobiology, climate science, and other interdisciplinary fields.
- EAR should take advantage of the broad appeal of field work, its modest cost, and its ability to capture the enthusiasm and research effort across a wide range of institutions by providing sufficient funding for graduate and undergraduate field work.

PARTNERSHIPS IN EARTH SCIENCE

Agency partnerships led by EAR will be essential for attaining many of the research objectives identified in this report. Well managed partnerships can foster broadly based research communities, leverage limited resources, and promote fruitful synergies. Cooperation with mission-oriented agencies can also be an effective mechanism for transferring NSF-sponsored basic research into practical applications. Geobiology, integrative studies of the Critical Zone, and paleoclimatology are obvious areas in which collaborations should be developed among a number of NSF divisions (e.g., the Atmospheric and Ocean Sciences Divisions, the Biological Sciences Directorate) and mission-oriented agencies (e.g., the U.S. Geological Survey, Department of Energy, National Oceanic and Atmospheric Administration, Environmental Protection Agency, and U.S. Department of Agriculture). The EarthScope project should also benefit from interagency cooperation on several levels: with the Ocean Sciences Division in gathering offshore data and linking to the Continental Margins Research program; with the Division of Civil and Mechanical Systems on earthquake research relevant to the Natural Earthquake Hazards Reduction Program and the Network for Earthquake Engineering Simulation; with the U.S. Geological Survey in deploying the Advanced National Seismic System; and with NASA in developing a satellite-based interferometric synthetic aperture radar system for observing active deformation. An effective initiative in planetary science will require careful coordination with NSF's Astronomical Science Division as well as with NASA. Improved core support for the study of Earth and planetary materials could be the basis for strengthening EAR's participation in the National Nanotechnology Initiative, and an EAR program on microorganisms in the environment should provide an appropriate Earth science focus for the NSF's cross-cutting program on Biocomplexity in the Environment. An ESNL program could solicit the cosponsorship of natural laboratories by other agencies, including state and local government agencies.

Continuing progress in Earth science will depend heavily on improvement to the computational infrastructure, including the development of community models that can function as virtual laboratories for the study of complex geosystems. EAR should be particularly aggressive in fostering substantive partnerships between Earth and computer scientists through the multiagency initiative on Information Technology for the Twenty-First Century and other programs.

REQUIRED RESOURCES

The committee's recommendations, taken together, lay out a basis for the manner in which the EAR Division can respond to major Earth science challenges and opportunities in the next decade. The committee estimates that the new funding needed to implement these recommendations would increase the EAR budget by about two-thirds. This increase would help to offset the recent decline in federal support of basic Earth science and would substantially strengthen the national effort in this important area of fundamental research.

1

Basic Earth Science and Society

INTRODUCTION

Human society is built upon a terrestrial foundation. Forests, farmlands, and cities are rooted in the Earth, and people draw sustenance from its outer layers in the form of water, food, minerals, and fuels. Earth science is thus a practical enterprise on which our society's survival depends. It is also a fundamental quest for understanding the natural world—an exploration to learn about the origin, evolution, and future of our planetary home. Curiosity about these basic issues sustains scientific inquiry even in areas where the utility of the research is less than obvious.

In fact, the fundamental and practical aspects of Earth science are intimately interwoven. Seismological and potential-field techniques developed for finding oil and minerals are now employed to image churning structures thousands of kilometers deep within the Earth's convecting mantle. These great thermal currents continually rejuvenate the face of Earth through plate tectonics, raising mountains, and causing earthquakes and volcanic eruptions. Helical motions even deeper, within the liquid outer core, generate the magnetic field that guides the compass and helps to shield the Earth's biosphere from solar and cosmic radiation. Efforts to simulate the deep-seated machinery of mantle convection and the core dynamo are stretching the limits of computing technology and providing a dynamical framework for synthesizing many previously disparate observations. On a more local scale, the methods of analytical geochemistry developed for the study of minerals, rocks, and soils have become powerful weapons in the fight against toxic pollution. Ultraprecise positioning techniques of space geodesy have measured the continental drift postulated by Wegener; they now monitor the accumulation of strain across dangerous faults such as California's San

Andreas, as well as maintain the reference grids used by land surveyors. Probing the Earth's past through a detailed reading of the geological record is furnishing information about the behavior of climate and ecological systems that will be crucial to a future in which human activities become ever more potent forces of global environmental change.

The linkage between basic and applied research is growing stronger because some of the toughest problems facing the United States and the world at large require a deep understanding of the physical, chemical, and biological processes that govern terrestrial systems. These practical issues cannot be addressed successfully without a vigorous program of basic research across the full spectrum of Earth science. Moreover, they call for substantial enhancements in the methodologies for integrating observations from the various disciplines into system-level models with predictive capabilities. Given the improved technical means for acquiring vast new data sets and modeling complex dynamic systems, the opportunities for furthering these aspects of the Earth science agenda have never been better.

Role of the National Science Foundation

Four federal departments and three independent federal agencies have significant activities in Earth science (Appendix A). These organizations support a mixture of basic and applied research, including multidisciplinary studies of mission-oriented problems ranging from environmental remediation and climate change assessment to anticipating the behavior of active faults and volcanoes. The National Science Foundation (NSF) plays a crucial role in this milieu as the sole agency whose primary mission is basic research and education. Only the NSF, through its Earth Science Division (EAR), provides significant funding for investigator-driven, fundamental research in all of the core disciplines of Earth science.[1]

The future of EAR is important because this NSF division now shoulders an increasing burden of the national effort in basic Earth science (Figure 1.1). In terms of buying power, the annual expenditures of EAR have grown about a factor of two during the last 20 years, reaching $97 million in 1999 (see Appendix A for a breakdown). However, the past few years have seen a substantial decline in the support of Earth science by other federal

[1]EAR is part of NSF's Geoscience Directorate, which also comprises the divisions of Atmospheric Science and Ocean Science. This report employs NSF terminology: *Earth science* is the subset of *geoscience* concerned with the study of the Earth's solid surface, crust, mantle, and core. The disciplines of Earth science include geology, geophysics, geochemistry, geobiology, hydrology, and related fields.

organizations. According to NSF's Federal Funds Survey, overall funding in basic Earth science fell from $555 million in 1993 to $388 million in 1997. When corrected for inflation, this amounts to a 37% reduction in federally supported basic research. EAR's share of basic research rose concomitantly, from 14% in 1993 to 24% in 1997. More than ever, Earth science in the United States depends on the ability of EAR to support research initiatives.

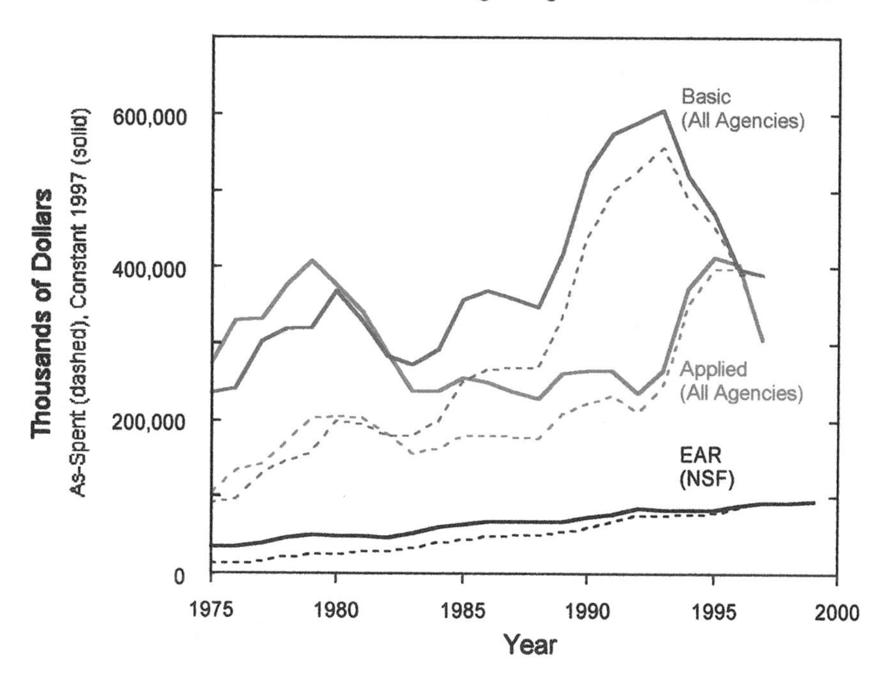

FIGURE 1.1 Total annual federal obligations for basic Earth science (blue lines) and applied Earth science (red lines), compared to the annual budget of NSF's Earth Science Division (black lines), in as-spent dollars (dashed lines) and constant 1997 dollars (solid lines). Since 1993, overall federal funding of Earth science has dropped substantially. SOURCE: Federal Funds Survey.

Organization of the Report

In this report, the Committee on Basic Research Opportunities in the Earth Sciences identifies areas of high-priority research within the purview of the Earth Science Division of NSF, assesses cross-disciplinary connections, and discusses the linkages between basic research and societal needs. Some general perspectives on these topics are given in this introductory chapter. The current EAR program is described in Appendix A. Chapter 2 ("Science Opportunities") surveys emerging fields of research and major scientific objectives that can now be addressed because of advances in theory, instrumentation, models, and data analysis. The committee's assessment of these science opportunities draws on the literature, workshop reports, direct experience, and letters from individuals. The names of committee correspondents and workshop reports are given in Appendix B. In Chapter 3 ("Findings and Recommendations"), the committee compares the science goals described in Chapter 2 with the stated objectives of EAR and suggests new programmatic directions where needed. The rationale for the initiatives is laid out in the context of the current NSF structure, with particular emphasis on potential interconnections, both within and outside the EAR programs.

APPLICATIONS OF BASIC EARTH SCIENCE TO NATIONAL PROBLEMS

NSF-sponsored basic research generates new understanding about the Earth that applies directly to national strategic needs. Basic research in Earth science affects human welfare in five major areas:

1. discovery, use, and conservation of natural resources—fuels, minerals, soils, water,
2. characterization and mitigation of natural hazards—earthquakes, floods and droughts, landslides, tsunamis, volcanoes,
3. geoscience-based engineering—urban development, agriculture, materials engineering,
4. stewardship of the environment—ecosystem management, adaptation to environmental changes, remediation, and moderation of adverse human effects, and
5. terrestrial surveillance for national security—arms control treaty verification, precise positioning, mapping, subsurface remote sensing.

Many of these strategic issues concern the near-surface environment, where interactions between rock, soil, air, water, and biota determine the availability of nearly every life-sustaining resource. This special interfacial region of mass and energy flux, which comprises terrestrial, lacustrine, and marine components of the uppermost continental crust, is here called the "Critical Zone." The Critical Zone is one of two primary loci of life on this planet and the environment for most human activity. The other major locus of life is the sea, but even there, the flux of nutrients from the Critical Zone is essential. Processes within the Critical Zone mediate the exchange of mass and energy; they are thus essential to biomass productivity, nutrient balance, chemical recycling, and water storage, and they ultimately determine the content of the geological record.

Natural Resources

The world's population reached 6 billion in 1999 and is increasing by more than 200,000 people per day. This populace requires food, fuels, raw materials, and water in ever-increasing quantities. Developing new technologies to deliver these resources depends on progress in many areas of Earth science.

Energy

Global energy consumption rose from 344 quadrillion British thermal units (BTUs) in 1990 to 376 quadrillion BTUs in 1995. The Department of Energy projects that this total will grow by an average of 2.1% per year until 2020. The reliance on fossil fuels is not likely to change appreciably during the next decade.[2] Oil consumption is expected to grow at an annualized rate of 1.8%, while natural gas usage will rise at 3.3%, faster than any other primary energy source. Worldwide, the burning of coal is likely to increase nearly as rapidly (3.0%), driven by energy demands in China, India, and other Asian countries.[3]

Historically, most of the research and development (R&D) related to the extraction of fossil fuels has been supported by the petroleum and mining industries, rather than by federal agencies. A worldwide oil glut that began in 1983 triggered a restructuring of the petroleum industry, reducing the industry's investments in basic research and its demand for Earth science professionals. Between 1991 and 1995, the petroleum refining and extraction industry reported cutbacks of 29% in R&D expenditures. The recent spate of

---

[2]The largest single energy source is oil (39%), which has increased in production from 57 million barrels in 1983 to 73 million in 1998. At present, fossil fuels account for about nine-tenths of the world's primary energy budget; only 2% is hydroelectric and 9% is nuclear.

[3]BP/Amoco Statistical Review of World Energy, June 1999; DOE Report EIA-0484, 1999.

mergers among the major companies (BP-Amoco, Exxon-Mobil) is leading to further consolidations of the industry's research activities. Nevertheless, the petroleum industry remains the largest employer of Earth scientists.[4] The short-term restrictions in the hiring of new Earth scientists have skewed the demographics, so that the petroleum companies will face problems in rejuvenating their professional science staffs. In the United States, this rejuvenation will depend heavily on graduates from research programs in Earth science supported by EAR.

Reservoir modeling, at the heart of modern optimization of oil and gas production, illustrates the link between basic and applied research. It involves the detailed characterization of hydrocarbon reservoirs and fluid properties at depth, and the use of large numerical simulations of multiphase fluid-flow to integrate many types of measurements (e.g., characterization of deep strata by three-dimensional seismic imaging and in situ well logging as a function of time) into predictive models of reservoir performance. Basic studies of fluid-rock interactions within petroleum reservoirs have much in common with investigations of a wide variety of hydrological, magmatic, and metamorphic systems in the upper crust. Given the commonality of processes among the various fluid reservoirs in the Critical Zone, the nation's energy industry has a major stake in the basic research sponsored by EAR.

Minerals and Other Raw Materials

The worldwide consumption of non-food, non-fuel raw materials was about 10 billion tons in 1995, almost double the amount in 1970. About 62% of these goods were construction materials such as crushed stone, sand, and gravel; 16% were industrial minerals; 7% metals; 6% nonrenewable organics; and 9% agricultural and forestry products.[5] The search for mineral resources is a geological activity, and the results have been spectacularly successful. Improvements in the techniques for finding and exploiting these natural resources have combined with the efficiency of global markets to make shortages in strategic materials rare. Earth science has been instrumental in transforming once-scarce raw materials into readily available, low-cost commodities.

A wide range of Earth science research—from the basic chemistry and physics of mass transfer to the global tectonic framework—has contributed

[4]According to the NSF's 1993 National Survey of College Graduates, 25% of all Earth science professionals and more than 50% of the professionals in the commercial sector were employed by the petroleum industry. See http://www.agiweb.org/career/geosec. html.

[5]USGS Fact Sheet FS-068-98.

to the development of geological models of ore-forming systems. Recent discoveries of significant ores have been strongly influenced by a basic understanding of Earth processes, including diamonds (mantle petrology), magmatic nickel (large igneous provinces), copper deposits (fluid-rock interaction), volcanic-hosted massive sulfide deposits (seafloor hydrothermal systems), and sediment-hosted lead-zinc deposits (basin-scale hydrology). Future discoveries may result from a better understanding of the physical, chemical, and biological processes involved in the formation and preservation of ore systems. Because the spatial and temporal distribution of mineral resources is highly variable, a key problem is to determine what controls the distribution of ore-grade mineralization. Large deposits of tungsten-tin and copper (Figure 1.2) are associated with periods of global arc volcanism, for example, but it remains unclear whether special conditions were necessary to create large upper crustal magma chambers or to trigger fluid release required to produce the ore. Research is also needed to determine the role of microbes in the modification and dispersion of ores in magmatic systems and associated hydrothermal environments, and the biological influences on the formation of sedimentary iron and uranium. Detailed, atomic-scale investigations of minerals, fluids, and mineral-fluid interfaces play a central role in this domain. However, with the continued success of science-based exploration, perhaps the greatest challenges for Earth scientists are not to find more resources—although this will surely remain important—but to produce them more efficiently and safely, to mitigate long-term environmental impacts, and to provide a scientific basis for long-term land-use decisions.

Water

According to the United Nations Environment Program, more than one-third of all people are without a safe water supply, and one-quarter will suffer from chronic water shortages during the next decade. By 2025, it is projected that 15 countries worldwide will have encountered water stress (i.e., consumption levels exceeding 20% of available supply), 9 will suffer from water scarcity, and 22 will have run up against a "water barrier" to further development.[6]

Informed decision making on water resources requires knowledge of the complex hydrologic systems operating within the Critical Zone and a predictive understanding of how they respond to natural and human modifications. At the Science and Technology Center (STC) for Sustainability of Semi-Arid

[6]Population change-natural resources-environment linkages in the Arab states region. Food and Agriculture Organization, 1996 (http://www.fao.org).

FIGURE 1.2 Mining of porphyry copper deposits, such as the Morenci, Arizona, deposit shown above, produces most of the world's copper from ancient subvolcanic intrusive centers. The volume of material moved by global copper production is comparable to the amount of material produced by arc volcanism—1 to 2 km^3 per year. This deposit represents approximately one-tenth of the time-averaged global arc magmatism and has supplied one-twentieth of the global copper. *Left*: The weathering profile of this early Tertiary subvolcanic system consists of a red hematite-rich zone leached of copper and sulfides, overlying light-colored sulfide-bearing copper-enriched zones. This profile likely developed during the Miocene through mineral-fluid reactions mediated by sulfur-metabolizing bacteria. *Right*: Transmission electron microscopy photo showing *Thiobacillus ferrooxidans* (circular features) in jarosite (dark material) formed during modern oxidation of sulfides at Morenci. Supergene chalcocite from the enriched ores contains similar fossilized bacteria. SOURCE: M.S. Enders, University of Arizona.

Hydrology and Riparian Areas, EAR's newest STC, the University of Arizona and its partners[7] seek to move scientific knowledge about these issues from research groups to the agencies responsible for managing water resources. The storage and flux of water in the Critical Zone are multidisciplinary problems that connect hydrology to the study of the oceans and atmosphere and to the solid Earth. On the geological side, the investigation of aquifers and groundwater systems now relies extensively on the use of geochemical and geobiological techniques, as well as geophysical methods. Such techniques have been used to investigate the cause and distribution of elevated levels of arsenic in groundwater, for example, which pose a serious health hazard in many parts of the world. The problem is particularly acute in Bangladesh, where groundwater provides 97% of the drinking water supply. Because contamination is localized, data from hydrogeological (groundwater flow velocities and directions), geophysical (resistivity, seismic), and geochemical (isotopic, trace element chemistry) techniques will help guide the placement of new wells that draw water with acceptable concentrations of arsenic.[8]

Soil

Soils are an immense and valuable natural resource. In their most obvious capacity, they serve as the foundation and primary reservoir of nutrients for agriculture and the ecosystems that produce renewable natural resources, but soils are also fundamental for waste disposal and water filtration, and as raw materials for construction and manufacturing activities. More generally, these biologically active, intricately structured, porous media—collectively called the pedosphere—mediate most of the life-sustaining interactions among the land, its surface waters, and the atmosphere. Organic carbon is recycled to the atmosphere through soils; about 25% of atmospheric carbon dioxide comes from soil biological oxidation reactions in the pedosphere, which contains twice as much carbon as the atmosphere and up to three times the carbon in all vegetation. Soils have a major influence on the hydrologic cycle.

---

[7]STC participants include the University of Arizona; Arizona State University; Scripps Institution of Oceanography; U.S. Geological Survey; University of New Mexico; Pennsylvania State University; Los Alamos National Laboratory; U.S. Department of Agriculture; Instituto del Medio Ambiente y el Desarrollo Sustentable del estado de Sonora; Desert Research Institute; Columbia University's Biosphere 2 Center; University of California, Riverside; Northern Arizona University; U.S. Army Corps of Engineers; and the University of California, Los Angeles.

[8]See West Bengal and Bangladesh Arsenic Information Centre (http://bicn.com/acic) and World Health Organization Fact Sheet 210 (http://www.who.int/inf-fs/en/fact210.html).

The water most people use comes from groundwater, streams, and lakes; regardless of its pathway, water quality is determined largely by the soils it passes through.

Soil management is thus crucial to sustaining and improving the human habitat, and issues related to land use, soil quality, degradation, and contamination now figure prominently in most policy decisions germane to the Critical Zone. The attributes of soil, coupled with climate variables, have traditionally been used in agriculture to predict the potential and limitations of land areas to produce food, feed, or fiber, but these same concepts are now being applied to all types of ecosystems. Moreover, the application of high-input farming, especially on marginally arable lands, has accentuated the environmental problems related to soil erosion, soil degradation through acidification, accumulation of toxic elements and salinization, and downstream contamination of aquatic systems from agricultural runoff. In the United States, precision agriculture with site-specific management is being practiced extensively to counter detrimental effects, but much is yet to be learned.

Judicious soil management will require increasing investments in soil science, including research on the fundamental physical, chemical, and biological processes involved in soil development. This type of basic research fits very well into the larger agenda of NSF-sponsored Earth science. As a geological process, soil development demonstrates the power of weathering, which in turn is a key process for issues as diverse as the availability of nutrients, the emission and capture of greenhouse gases, the chemistry of the ocean, and the evolution and longevity of landscapes. How water flows through soil and interacts chemically with the pedosphere is fundamental to hydrology and climatology. Microbial processes in soil are a primary topic for novel research in geobiology, and the microscopic structure of soil is a new focus in the study of Earth materials. The opportunities to connect the study of soils to other aspects of geoscience through the basic research programs of EAR are therefore expanding.

Natural Hazards

The Critical Zone in which humans and many other biota live is a high-energy, often dangerous, interface. Here the solar-powered processes in the Earth's fluid envelope interact with the tectonic processes powered by heat escaping from its deep interior. The atmosphere transports water from the oceans to the continents, where irregular patterns of rainfall and evaporation combine with the complex hydrological response of the land surface and its vegetation to produce a chaotic sequence of flooding and drought. The planetary heat flux drives plate tectonics and melts rock to form the magmas that erupt in volcanoes. The plate motions accumulate stresses in the brittle part of the lithosphere, releasing strain energy through sudden failures on faults, causing earthquakes. Plate tectonics pushes up mountains and creates other topographic features, which release gravitational energy in the form of landslides and avalanches. When major landslides and earthquakes occur under the ocean, some of the potential energy that is released can propagate in the form of huge sea waves (tsunamis), inundating coastlines thousands of miles away.

Floods, droughts, severe storms, volcanic eruptions, earthquakes, landslides, and tsunamis compose a catalog of natural disasters that have wreaked destruction since the beginning of civilization. Only recently, however, has the changing nature of these threats been recognized. The process of urbanization begun in the Industrial Revolution continues apace; in 1950, only 3 out of 10 people lived in urban areas, while by 2030 this fraction will nearly double. As populations and the fragile infrastructures on which they depend concentrate in large urban areas, the risks of natural hazards, especially the economic risks, grow correspondingly. With regard to seismic hazard, Japan is fairly well prepared for earthquakes. Yet the modest-sized earthquake (magnitude 6.9) that struck Kobe on January 18, 1995, killed 5500 people and resulted in an economic loss of nearly $200 billion. According to one recent study, a repeat of the great 1923 Kanto earthquake (magnitude 7.9) would devastate modern Tokyo: the direct economic losses would total a staggering $2.1 trillion to 3.3 trillion, equivalent to 44-70% of Japan's annual gross domestic product.[9] An event of this magnitude clearly would have an impact extending well beyond any one nation, affecting the entire global economy and thereby directly influencing the welfare and security of the United States.

On a worldwide basis, the problem of urban hazards is further amplified by the fact that the most severe natural disasters—earthquakes, hurricanes, typhoons, and volcanic eruptions—tend to be concentrated in low-latitude, coastal regions, where ambient environmental conditions support large populations and the current economic development is most intense.[10]

---

[9]Risk Management Solutions, Inc., Menlo Park, California, *What If the 1923 Earthquake Strikes Again? A Five-Prefecture Tokyo Region Scenario*, 97 pp., November 1995. Another report in the same series estimated that repeat of the 1906 San Francisco earthquake, also magnitude 7.9, would result in a 3000-8000 deaths and a direct economic loss of $170 billion to $225 billion (1994 dollars).

[10]R. Bilham, in *Reduction and Predictability of Natural Disasters*, J. Rundle, D. Turcotte, and W. Klein, eds., Santa Fe Institute Studies in the Science of Complexity, Vol. XXV, Addison-Wesley, pp. 19-31, 1996. Earthquakes and volcanoes occur predominately on or near the boundaries of the tectonic plates; the most active plate boundaries are concentrated at low latitudes, probably because the Earth's principal moment of inertia is determined by convective upwellings and downwellings.

Applied research on different types of natural hazards has a common set of practical objectives:

- education of the general populace about the threat of natural hazards and the steps that can be taken to reduce risks (public preparation),
- quantification of what will happen during a particular event scenario (deterministic hazard analysis) or over an ensemble of possible events (probabilistic hazard analysis)—both are used by engineers to reduce human casualties (safety engineering) and economic losses (performance-based engineering).
- specification of individual disasters in terms of where, when, and how large (event prediction)—in some cases, such as earthquakes and tsunamis, an accurate prediction of occurrence times may not be possible, in which case the objective of "when" may be weakened to "how frequently" (long-term forecasting), and
- accurate and timely information about the occurrence and circumstances of disastrous events for public notification (including early warning, when possible) and for use by government officials and emergency management personnel (rapid response).

Experience shows that these practical objectives are difficult to attain without precise observations and good understanding of the phenomena involved in natural hazards. Because the processes of magma motion through the upper crust can be detected and analyzed, the most dangerous volcanic eruptions can be predicted in a useful way. The 1991 eruption of Mt. Pinatubo in the Philippines was forecast by the U.S. Geological Survey (USGS) early enough to allow the region to be evacuated and aircraft and other equipment to be removed from Clark Air Force Base. The forecasts are estimated to have saved 5000 to 20,000 lives and economic losses of at least $200 million.[11] On the other hand, the most dangerous earthquakes cannot be predicted deterministically, because no reliable, precursory indicators of their timing, location, and magnitude have been discovered. Is this because such precursors do not exist or because the fault-rupture process is too poorly understood to know which kinds of behavior are most diagnostic of a large, impending event? These questions remain the subject of vigorous basic research, supported by EAR as part of NSF's participation in the National Earthquake Hazard Reduction Program.

[11]USGS Fact Sheet 115-97.

Geoscience-Based Engineering

The role Earth science plays in civil and environmental engineering is often underappreciated. The foundations of cities, as well as the transportation networks that connect them and the lifelines of utilities that sustain them, are necessarily embedded in the Earth. The geology of the Earth's surface therefore determines many aspects of how this development takes place, from the distribution of skyscrapers on Manhattan Island to the layout of freeways across the Los Angeles basin, both of which reflect fundamental geological structures. The geological problems of the "built environment" are the subject of geotechnical engineering, which is a major consumer of Earth science information. Geotechnical engineers must understand the short-term and long-term properties of soils, rocks, groundwater movement and composite geological formations, including their static strengths, responses to dynamic stresses and deformations, and degradation by weathering and other alterations. As urban areas develop and expand, these engineers are being called upon more frequently to slow down natural processes or even stop them from running their course. The challenges cataloged in one popular account[12] include preventing the Atchafalaya River from capturing the Mississippi River to keep it flowing past New Orleans, stopping the encroachment of basaltic lava on a community in Iceland, and minimizing the impact of flooding and debris flows from the steep scarp of the San Gabriel Mountains adjacent to Los Angeles. Basic research furnishes the knowledge of Earth materials and processes needed to address these engineering challenges.

The study of Earth materials has contributed significantly to materials science and engineering. For example, the study of nanocrystals and biomaterials originated in research on soils and biominerals, and Earth scientists have pioneered the development of substances ranging from high-temperature superconductors to superhard materials. Moreover, they have been leaders in the development of new analytical technologies, ranging from ultraprecise isotopic measurements and atomic resolution imaging of minute particles to the application of synchrotron, neutron, and other major facilities to the study of complex natural substances. These sophisticated analytical and experimental techniques have been employed in applications ranging from the engineering of synthetics to novel methods for forensic investigations. Earth scientists have also led in research at ultrahigh pressures, with applications to the physics and chemistry of materials as well as to simulating the deep interiors of planets.

---

[12] J. McPhee, *The Control of Nature*. The Noonday Press, NY, 272 pp., 1993.

The Environment

With the increase in world population, the Earth has seemingly shrunk from a vast realm of unspoiled lands and seas to a small blue planet with limited resources for sustaining human activities. Environmental concerns now factor into policy decisions on the extraction and production of energy and mineral resources, as well as the disposal of their waste products. A key environmental issue is the storage of radioactive products from the nuclear arms program of the Cold War, spent fuel rods from nuclear power plants, and high-level radioactive wastes from medicine and other applications. Continued research on the level of seismic and volcanic activity and the stability of the water table will be needed to ensure the isolation of long-lived radioactive waste in designated underground repositories such as Yucca Mountain. Materials studies, including at the microscopic scale, have played an important role in revealing the potential for containment within or leakage through both the engineered barriers and the natural media of prospective waste disposal sites.

Mining, milling, and in situ leaching also produce wastes that are toxic to living organisms. Hardrock mining of metalliferous deposits can release metals and chemicals such as cyanide to surface and groundwater and to aquatic ecosystems. Moreover, mining activities directly affect terrestrial ecosystems because they destroy habitat and alter migration patterns by creating barriers and fragmenting animal territories. Geochemical, hydrological, mineralogical, and microbiological research and monitoring will be key to mitigating these adverse effects. In dealing with these and other challenging problems of the near-surface environment, basic research is needed to achieve an understanding that reaches from the atomic scale, at which the detailed fluid-flow patterns and the distribution of contaminants between fluids and solids are determined, to the scale of major geological features and hydrologic systems, which govern the regional containment and dispersal of contaminants.

On an even larger scale, human activities are now capable of causing substantial, if unintended, global environmental change. Anthropogenic contributions to rising atmospheric CO_2 and other greenhouse gas concentrations and their potential impact on future climate are issues of global economic and political significance. Earth science is playing a significant role in understanding the global carbon budget and key aspects of greenhouse forcing, not only in the present, but at longer time scales, which are accessible only through the geological record. Information on paleoenvironments, which is extensive but relatively unexploited, can be used to identify forcing factors that have controlled climate in the past, their variation over time, and the causes of rapid transitions in the climate state. In particular, geochemical, isotopic, and paleontologic analyses of marine sediments and fossils can be used to infer the extent of glaciers and ice sheets, as well as the temperature and composition of the oceans. The abundance and range of species, the nature of the land cover, and the position of shorelines can be inferred from analysis of terrestrial sediments, soils, and fossils. Such research provides important constraints on climate models, which are extending increasingly into the geological past.

National Defense and Global Security

Earth science has found significant applications in the arenas of national defense and global security. A range of technologies based on Earth science are essential components in the global monitoring and verification of nuclear test bans, nuclear nonproliferation treaties, and other arms control measures. From the first multinational discussions in Geneva in the late 1950s, it has been recognized that reliable identification of small underground nuclear explosions is the primary technical issue confronting the verification of a comprehensive nuclear test ban treaty (CTBT). The resulting U.S. program in nuclear explosion seismology stimulated basic as well as applied research. To improve seismological detection and location capabilities, for example, the Department of Defense developed a 120-station World Wide Standardized Seismographic Network (WWSSN). The WWSSN sharpened the images of global seismicity, which played a significant role in the discovery of plate tectonics and yielded much better models of deep-Earth structure that in turn resulted in a better understanding of the Earth's composition and internal dynamics.

After decades of negotiations, the CTBT was opened for signature on September 24, 1996, and has thus far been endorsed by 154 nations. Adherence to the treaty will be verified by the International Monitoring System (IMS). The IMS is a worldwide distribution of permanent stations designed to detect clandestine nuclear explosions by measuring wind-transported radionuclides and waves transmitted through the atmosphere (infrasound), oceans (hydroacoustic), and solid Earth (seismic). Basic research related to the hydroacoustic and infrasound networks will open new avenues of research in the Earth sciences (e.g., the infrasound network can be used to count meteorites). EAR-sponsored research related to the CTBT focuses primarily on seismology, although radionuclide and other geochemistry studies are also critical for verification of the CTBT and the Chemical Weapons Convention of 1993. Recent advances in particulate and isotopic analyses have considerable potential for improving the capabilities in radionuclide monitoring.

Earth science contributes to national security in a number of other ways. Precise geodetic measurements of the Earth's topography, gravity field, and the active deformation of its solid surface are crucial to military as well as civilian navigation. The electromagnetic properties of the solid Earth must be studied to determine their effects on global communications. On a much smaller spatial scale, electromagnetic sounding methods are employed by the military to detect unexploded ordnance. Geophysical remote sensing is used to gather intelligence on subsurface operations that require tunneling and other excavations.

THE AGENDA FOR BASIC RESEARCH

The study of the Earth remains a true science of discovery. From the theory of evolution to the theory of plate tectonics, breakthroughs in this field have influenced deeply our thinking about the natural world, and there is every reason to believe that discoveries of similar significance will be made in the future, especially about events and processes still obscured in the Earth's past or hidden at depths within its interior. Many great unsolved problems spring easily to mind: the origin of life; causes of rapid biological diversifications and extinctions; early evolution of the solar system and planetary accretion; segregation of the core, inner core and continents; workings of active fault systems; mechanisms of climate transitions; and extent of the deep biosphere. At the same time, it is important to recognize that scientific discoveries are not born in isolation, but usually arise in a context prepared by the continuing integration of new data into better models of how the world works. In Earth science, the rate of this synthesis has been accelerated by major improvements in three types of research capabilities: (1) techniques for deciphering the geological record of terrestrial change and extreme events, (2) facilities for observing active processes in the present-day Earth, and (3) computational technologies for realistic simulations of dynamic geosystems. Exploiting these capabilities and extending their range offer a new agenda for basic research.

Reading the Record of Terrestrial Change and Extreme Events

A distinguishing feature of Earth science is its access to the planet's unique history "written in stone." This geological record comprises a wealth of information about terrestrial and extraterrestrial events and conditions, from the present back into the farthest reaches of time. It is preserved in the rocks and fossils of the continents, their margins, and the deep seafloor, as well as in a wide range of extraterrestrial materials collected in the form of meteorites, cosmic dust, and samples ferried by spacecraft from other bodies in the solar system.[13] Sequences of sedimentary rocks record events on time scales ranging from the subannual to billions of years. Metamorphic rocks found in ancient continental regions yield radiometric ages of up to 4 billion years, and they document processes active when the Earth was still fresh from a Hadean period of planetary formation and bombardment. Meteorites extend this record back to the earliest events in the condensation of the solar nebula, 4.56 billion years ago, and minute samples of cosmic dust have been identified that push this chronicle even further back to the actual manufacturing of the chemical elements in earlier generations of stars.

The methods available for reconstructing the history of the Earth and its parent nebula have been greatly extended and improved. Entirely new techniques are now available from previously inaccessible isotopic systems; examples include the use of tungsten isotopes to constrain the timing of the segregation of the Earth's metal core and osmium isotopes to date refractory mantle samples. Augmented capabilities have also come from substantial improvements to old techniques, such as recalibration of the carbon-14 method, and the use of accelerator mass spectrometry to extend its temporal resolution and reduce the requisite sample size.

State-of-the art analytical techniques promise to define much more precisely the timing, duration, and lateral extent of "extreme events," which include major magmatic eruptions, large bolide impacts, unusual excursions in global climate, and collapses and reversals of the Earth's magnetic field. During these rare occurrences, conditions at the Earth's surface have greatly exceeded their usual range, and they have therefore exerted a disproportionate influence on the evolution of the planet and its biosphere. Absolute dates on the ages of individual units within geological formations can now be obtained from the uranium-lead and potassium-argon systems with sufficient precision (fractions of a percent) to estimate the duration of the Cambrian "explosion" (the sudden first appearance of macroscopic, skeleton-bearing life), the great mass extinction at the Permo-Triassic boundary, and the huge outpourings of magma (millions of cubic kilometers) in the form of flood basalts that have occurred at irregular intervals throughout Earth history.

Extreme events of short duration can be difficult to decipher because only a small fraction of history is preserved in the geological record. What happened must be inferred from fragmentary evidence, such as the sequence of sedimentary deposits (floods, mudflows, major storms), juxtapositioning

---

[13]Almost all of the techniques for deciphering and interpreting extraterrestrial samples are common to the investigation of geological materials, and most NSF support for these activities falls under the Earth Science Division. For the purposes of this report, the disciplinary domains of planetary science and cosmochemistry are considered an integral part of Earth science.

of different paleoenvironments (earthquakes), or characteristic mineralogy or chemistry (bolide impacts). On the other hand, extreme events of longer duration can be easier to recognize than smaller, more transient changes. Climatic extremes such as the "hothouse" conditions of the Cretaceous resulted in widespread bauxite deposits (lateritic weathering), petroleum generation (high marine stands and biologic productivity), and iron- and phosphorous-rich sedimentary rocks (marine upwelling). Precambrian mantle plumes have been postulated as the cause of the major magmatic eruptions that led to the formation of large igneous provinces and their associated mineral deposits, which are themselves chemical extremes.

Similarly, relatively recent extreme events are recognizable in sediment and ice cores. Analysis of such cores has led to a greater appreciation of the transient nature of climate change and the potential for abrupt and amplified responses following small perturbations in atmospheric and oceanic processes. For example, marine records from the Paleocene-Eocene boundary reveal an intense warming period, lasting no more than 10,000 years, associated with a large benthic extinction, changes in ocean circulation patterns, and increased ocean temperatures, especially in higher latitudes. Isotopic analysis suggests that a catastrophic release of methane, possibly from marine clathrates, resulted in a sharp increase in greenhouse warming. On land, soil and vertebrate fossil evidence links this event to the sudden onset of warm terrestrial climates and the first appearance of modern mammal lineages.

The combination of precise geochronology and systematic field investigations can reconstruct surprisingly complete and detailed accounts of what happened during major events in Earth history. One of the most fascinating detective stories in all of geoscience is the discovery of the Chicxulub crater off the Yucatan peninsula, the "smoking gun" that confirmed the Alvarez hypothesis that a bolide impact killed the dinosaurs at the end of the Cretaceous.[14] Research of this type is yielding an increasingly rich picture of Earth processes, which in turn is helping to assess the significance of future changes and extremes, the mechanisms that might trigger them, and the hazards to human life that could result.

Observing the Active Earth

Until recently, only rudimentary instrumentation was available for collecting synoptic data on global processes in real time, and the monitoring

---

[14]W. Alvarez, *T. Rex and the Crater of Doom*. Princeton University Press, Princeton, N.J., 236 pp., 1997.

of processes on regional and local scales was spotty at best. However, the ongoing "digital revolution" has significantly improved the observational capabilities of Earth science through the development of many new remote-sensing and direct-sampling technologies. In addition, data-gathering efforts have been greatly facilitated by worldwide communication systems that can transmit high-resolution observations of many variables from remote locations in real or near-real time. Newly available technologies range from space-based platforms and global networks of surface observatories to extremely sensitive instruments that can measure Earth materials and processes in both the laboratory and the field.

Laser altimetry from aircraft can be combined with accurate digital elevation models to investigate the surficial processes of erosion and sedimentation at the meter scale. Interferometric synthetic aperture radar aboard satellites can map decimeter-level deformations of fault ruptures, magma inflation of volcanoes, and ground subsidence continuously over areas tens to hundreds of kilometers wide. These images of the strain field complement the even more precise, pointwise measurements from the satellite-based Global Positioning System (GPS). GPS receivers can be located with millimeter precision over baselines of thousands of kilometers and can thus be used to map long-term strain rates across wide plate boundaries, such as in the western United States, while arrays of GPS stations can be used to measure the short-term deformations associated with volcanoes and earthquakes.

Observing processes that are active beneath the solid surface is particularly challenging, because the Earth's interior is inaccessible and characterized by extreme conditions. Rocks in the outer few kilometers of the crust can be sampled directly by trenching, tunneling, and drilling. Trenching to a depth of a few meters provides a means of studying environments (e.g., paleosols, fluvial systems) and processes (e.g., weathering, faulting) that operated in the relatively recent geological past. Paleoseismologists have made particularly effective use of trenching techniques to discover and date precisely the history of individual large earthquakes on major faults. Although laborious and expensive, drilling is often the best method for probing more deeply buried rock masses and collecting in situ measurements of active geological processes. Novel logging techniques developed by the petroleum industry, such as nuclear magnetic resonance and electromagnetic borehole imaging, are furnishing unparalleled data on the environments deep within sedimentary basins and continental basement rocks. The German KTB drilling project, which penetrated to a depth of 9100 m in 1994, furnished key insights into crustal processes, revealing the near-critical state of crustal stress predicted by Byerly's relationship, which has important implications for earthquake mechanics. In addition, the project confirmed hydrostatic pore pressure at great depth in the Earth's crust and detailed geochemical data provided constraints on how the hydrological cycle operates. Samples from more recent deep drilling of a Hawaiian volcano are yielding a better understanding of the physics and chemistry of deep-seated sources of volcanism. Deep drilling of active faults such as the San Andreas offers great promise for elucidating earthquake processes, including the role of fluids in fault mechanics.

Earthquakes and controlled (artificial) seismic sources generate a variety of elastic waves that encode an immense amount of information about the Earth through which they propagate. This illumination can be captured on arrays of seismic sensors and digitally processed into three-dimensional images of Earth structure and moving pictures of earthquake ruptures. Seismology thus gives geoscientists the eyes to observe fundamental processes within the planetary interior. In the past 15 years, the NSF-funded Incorporated Research Institutions for Seismology (IRIS) have provided new tools of seismological imaging to a broadly based user community, and the results have transformed the study of the solid Earth. The USArray program, a part of the recently proposed EarthScope initiative, would greatly extend these seismological capabilities, allowing the Earth beneath North America to be imaged at much higher resolution than with existing instrumentation (see Chapter 2, Box 2.2).

Measurements of the gravitational potential and electromagnetic fields constrain the mass variations and electrical properties of the interior, information that is complementary to seismological imaging and critical to its interpretation in terms of dynamic processes. The upcoming Gravity Recovery and Climate Experiment (GRACE) mission, for example, will measure the time-dependent component of gravity and thereby significantly advance studies of postglacial rebound, structure, and evolution of the crust and lithosphere; the hydrologic cycle; and mantle dynamics and plumes.[15]

Geophysical interpretation also requires the understanding of Earth materials at the extreme pressures, high temperatures, and special chemical environments of the deep interior. Only by characterizing materials at these conditions can one translate the observations of remote sensing and geochemical sampling into a concrete understanding of the current state, evolution, and ultimately, origin of the planet. State-of-the-art laboratories and apparatus are needed for this purpose. Both static and dynamic (shock-wave) methods are employed, with the NSF Center for High-Pressure Research (CHiPR) being a leader for the United States in the application of synchrotron facilities to static high-pressure experiments with diamond-anvil cells and multianvil presses. High-pressure techniques developed for the study of the deep interiors of the Earth and other planets now allow the

[15] *Satellite Gravity and the Geosphere: Contributions to the Study of the Solid Earth and Its Fluid Envelope*, National Academy Press, 112 pp., 1997.

densities of solids and fluids to be changed by as much as an order of magnitude in the laboratory, revealing unforeseen properties while also producing new types of materials and supplying novel insights into condensed-matter physics (Figure 1.3).[16]

The chemical and isotopic composition of volcanic rocks on the Earth's surface and of accidental inclusions of mantle rocks (xenoliths) in such lavas carries vast amounts of information on magma formation processes and the history responsible for the Earth's chemically layered structure. The petrological and geochemical signatures of some xenoliths indicate that they have come from the midmantle transition zone and even the lower mantle, providing samples of the deep interior that can be studied directly in the laboratory. Advances in geochemical techniques and interpretation continue to expand and refine the understanding of chemical phenomena occurring within the mantle and core, both at the present time and in the distant geological past. Of particular interest is the increasing convergence between geochemical and geophysical approaches and the brightening prospects for a unified model of deep-Earth dynamics.

Measurement systems are now providing data on active Earth processes that are of unprecedented quality and quantity. To take full advantage of these rich sources of information, geoscientists will have to harness the power of advancing information technologies to collect and assemble raw data, to process and archive data products, and to make these products widely available to researchers and other users. A number of challenges can be identified: how to collect data in real time at modest cost from expanding global networks of sensors, many in remote locations; how to reconfigure networks for robust operation when components fail, emergencies arise, or demands peak; how to ensure prompt delivery of data to users with time-critical needs (e.g., rapid response to natural disasters) while maintaining quality control and accessibility by lower-priority users; how to process heterogeneous data streams quickly enough that the data volume does not overwhelm managers and users; how to archive data in a way that enhances research capabilities rather than leading to overfull data warehouses. This type of information management will require innovations in Internet connectivity, multimedia information processing, digital libraries, and visualization techniques. Geoscientists are in an excellent position to exploit and contribute to the research being done in all of these areas by the information technology communities.

---

[16] R.J. Hemley and N.W. Ashcroft, The revealing role of pressure in the condensed matter sciences, *Physics Today*, v. 51, p. 26-32, 1998.

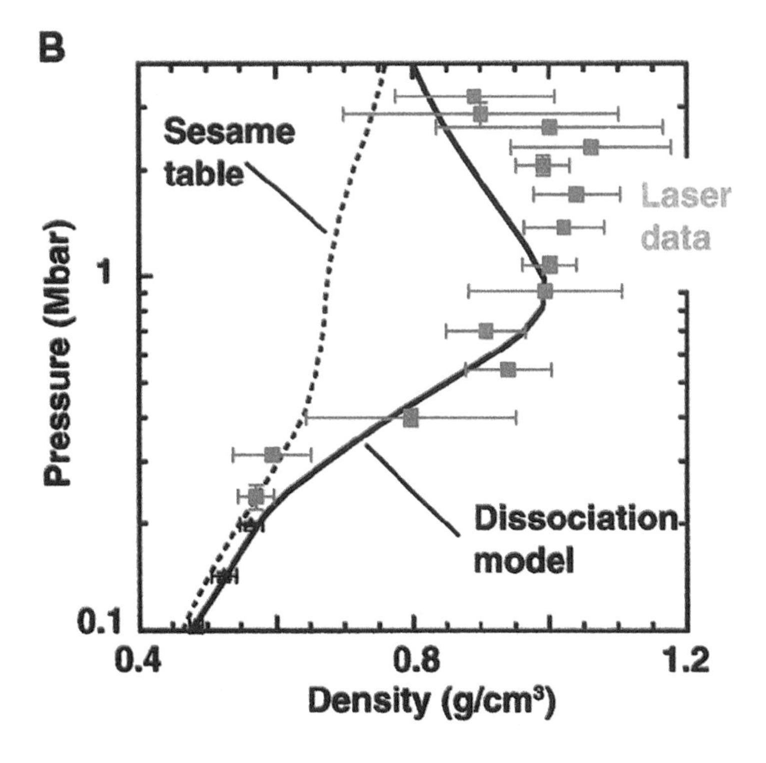

Figure 1.3 New experiments document the appearance of the dense metallic state of hydrogen at pressures greater than 0.6-1.0 Mbar (60-100 GPa), but with much higher densities than had previously been expected (dotted curve; the results shown here are for the deuterium isotope). Knowledge of the properties of hydrogen is needed for understanding the origin and evolution of planetary systems, with metallic hydrogen being the primary constituent of giant planets and of stars. SOURCE: Reprinted with permission from B.A. Remington, D. Arnett, R.P. Drake, and H. Takabe, Modeling astrophysical phenomena in the laboratory with intense lasers, *Science*, v. 284, p. 1488-1493, 1999. Copyright 1999 American Association for the Advancement of Science.

Modeling Geosystems

Much of Earth science concerns the operation and evolution of various terrestrial systems over scales that range from global (climate, mantle convection, core dynamo) to regional (orogens, active fault systems, sedimentary basins) to local (volcanoes, petroleum reservoirs, landforms, soils, aquifers). Better knowledge of these geosystems, especially improvements in predicting their behaviors, can be of immense practical value; thus, achieving a predictive understanding has become a major research goal. Because most geosystems involve many components that interact nonlinearly over a wide spectrum of spatial and temporal scales, the behaviors they display are not amenable to classical theoretical analysis and manual calculation. Indeed, because of the rapidly expanding capabilities for observing terrestrial activity, the data volumes on geosystems will soon be measured in petabytes. The interpretation of such vast quantities of data lies beyond the expertise and ability of the lone scientist, requiring collaborations among large groups of investigators from a variety of disciplines. The primary integrative mechanism for this multidisciplinary activity is the system-level model.

Only in the past few years have computational capabilities permitted numerical simulations of an interesting spectrum of geosystem behaviors in three spatial dimensions. For instance, although the first attempts to model solid-state convection in the Earth's mantle date from the early 1970s, the numerical resolution required to represent mantle convection properly in three dimensions was attained only in the 1990s. The first simulation of a self-sustaining core dynamo based on a realistic set of governing equations was not achieved until 1995. For some problems, such as simulations of active fault systems, full three-dimensional calculations over the appropriate scale range exceed the capabilities of even the largest available computers.

Continuing progress in geosystem modeling will depend heavily on improvements to the computational infrastructure of Earth science, including computational algorithms for exploiting parallel computers and other hardware, access to distributed computing and collaborative environments, advanced methods for code development and sharing, software libraries, visualization tools, and data management capabilities. The need for community models that can function as "virtual laboratories" for the study of particular geosystems presents a major challenge because new organizational structures will have to be set up to develop, verify, and maintain the requisite software components. The strategies and tools for this type of collaborative research are being developed by computer scientists in partnership with other research communities, and Earth scientists can learn and profit from participating in these efforts.

Most geosystems are so complex that the ability to extrapolate the observed behaviors into new regimes and confirm them with additional data becomes an essential measure of how well a system is understood. Predictions made from system-level models thus play an integral role in an iterated cycle of data gathering and analysis, hypothesis testing, and model improvement. The reliance on this type of model-based empiricism has significant implications for the organizational structure of geoscience, in addition to its epistemology, because it offers a framework for integrating observations from many disciplines. In the study of active faulting, for example, geologists map faults, geomorphologists date fault motion, seismologists locate earthquakes, geodesists measure deformations, and rock mechanists investigate the frictional properties of fault materials. Numerical simulations of active fault systems attempt to bring together these various types of observations in the context of a self-consistent model. The success of such a model in reconciling diverse types of information can thus be used to confirm the compatibility of the data from different disciplines and ferret out inconsistencies, in addition to giving researchers confidence in their underlying assumptions and hypotheses.

The problems of relating observations and simulations are particularly difficult in the research fields sponsored by EAR, because the solid Earth is characterized by physical and chemical processes that generally have shorter ranges and longer durations than the fluid systems investigated by meteorologists and oceanographers. Two representative examples illustrate this point. First, the global circulation time is on the order of a month for the troposphere and about a thousand years for the deep ocean, but it exceeds 100 million years for mantle convection. Second, chemical diffusion is an important process both above and below the Earth's solid surface, but the diffusivities of the common cations are 10 to 15 orders of magnitude lower in solid rock than in liquid water.

Fluid-bearing geosystems in the Critical Zone and upper crust—rock bodies containing magmas, petroleum, or water—present special challenges in this regard because the relevant processes range from the atomic level (i.e., sorption-desorption on mineral surfaces) to tens of kilometers or more. Their elucidation requires systematic, coordinated observations involving multiple disciplinary techniques that are spatially dense and extend over long time intervals. Field studies of this type are often most efficiently accomplished through the joint efforts of several groups of investigators in carefully chosen localities. Measurements and experiments within such "natural laboratories" may have to continue for many years. This mode of research is becoming more common as the trend toward the quantification of geological processes and system-level behaviors accelerates. Hence, the demand for basic research funds to invest in natural laboratories can be expected to increase.

2

Science Opportunities

This chapter focuses on some specific topics in Earth science that appear to be ripe for major breakthroughs during the next decade. Six major areas are discussed, roughly organized according to proximity and scale: (1) the near-surface environment or "Critical Zone," (2) geobiology, (3) Earth and planetary materials, (4) the continents, (5) the deep interior, and (6) the planets. The committee emphasizes that this is not intended to be a comprehensive list of exciting areas in Earth science, or to represent the full variety of research currently sponsored by the Earth Science Division (EAR) of the National Science Foundation (NSF), but rather to provide key examples of vigorous research areas from which some important programmatic directions can be discerned. The committee's findings and recommendations regarding these directions can be found in Chapter 3.

THE CRITICAL ZONE: EARTH'S NEAR-SURFACE ENVIRONMENT

The surface and near-surface environment sustains nearly all terrestrial life. The rapidly expanding needs of society give special urgency to understanding the processes that operate within this Critical Zone (Box 2.1, Figure 2.1). Population growth and industrialization are putting pressure on the development and sustainability of natural resources such as soil, water, and energy. Human activities are increasing the inventory of toxins in the air, water, and land, and are driving changes in climate and the associated water cycle. An increasing portion of the population is at risk from landslides, flooding, coastal erosion, and other natural hazards.

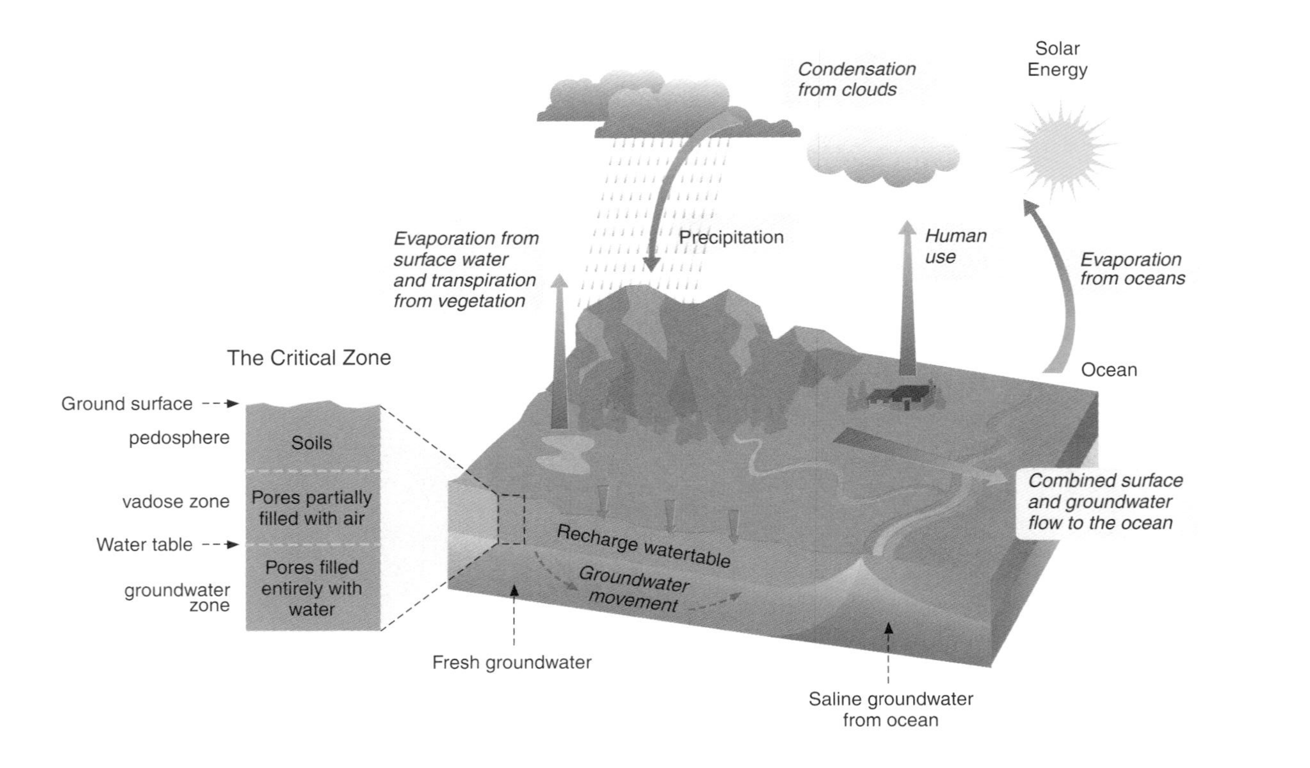

FIGURE 2.1 The Critical Zone includes the land surface and its canopy of vegetation, rivers, lakes, and shallow seas, and it extends through the pedosphere, unsaturated vadose zone, and saturated groundwater zone. Interactions at this interface between the solid Earth and its fluid envelopes determine the availability of nearly every life-sustaining resource.

The Critical Zone is a dynamic interface between the solid Earth and its fluid envelopes, governed by complex linkages and feedbacks among a vast range of physical, chemical, and biological processes. These processes can be organized into four main categories: (1) *tectonics* driven by energy in the mantle, which modifies the surface by magmatism, faulting, uplift, and subsidence; (2) *weathering* driven by the dynamics of the atmosphere and hydrosphere, which controls soil development, erosion, and the chemical mobilization of near-surface rocks; (3) *fluid transport* driven by pressure gradients, which shapes landscapes and redistributes materials; and (4) *biological activity* driven by the need for nutrients, which controls many aspects of the chemical cycling among soil, rock, air, and water.

Critical Zone processes are highly nonlinear and range across scales from atomic to global and from seconds to aeons (Figure 2.2). The scientific challenges are illustrated by the problem of methane flux from wetlands and sediments, which reflects microbial processes and chemical gradients on small scales; the influence of vegetation and nutrient inputs on regional scales; and climatic factors such as precipitation and temperature on global scales. The scientific requirements thus include the development of process models that can capture the scale dependence (or invariance) and reconcile observations from one scale to another.

Practical applications often rely on the predictive capability of such models. To engineer the safe disposal of radioactive wastes, geoscientists must be able to predict reliably the effects of hydrologic and geologic processes on underground disposal sites for thousands of years. The historical record of direct observations is far too short to capture the full range of possible behaviors in the Critical Zone, and extensive use of the geological record becomes necessary. For instance, biogeochemical cycles are studied over decades to centuries through high-precision geochemical analysis of ice and sediment cores and marine organisms, while the geologic forcing factors (e.g., volcanism, topography) are constrained through the analysis of sedimentary and volcanic rocks deposited over millions of years.

Science Opportunities

Processes in the Critical Zone control soil development, water quality and flow, and chemical cycling, and they regulate the occurrence of energy and mineral resources. A better understanding of Critical Zone is necessary to assess the impact of human activities on the Earth surface and to adapt to their consequences. It is not possible in this short report to do justice to all of the pressing scientific issues that bear on the near-surface environment, hence only some pertinent examples are given.

Box 2.1. The Critical Zone

The Critical Zone, depicted in Figure 2.1, comprises the outermost layers of the continental crust that are strongly affected by processes in the atmosphere, hydrosphere, and biosphere. The upper boundary of the Critical Zone includes the land surface and its canopy of lakes, rivers, and vegetation, as well as its shorelines and shallow marine environments. On land, the shallower part is the *vadose zone*, in which unconsolidated Earth materials are intermixed with soil, air, and water. This porous medium is a host for many chemical transformations mediated by radiant energy, atmospheric deposition, and biological activity. Its storage capacity influences runoff and groundwater recharge, affecting both the flow and the quality of surface and subsurface waters. Within the vadose zone is the *pedosphere*, a collective term for soils at the land surface. The characteristic layering of the soil profile reflects the strong interaction of climate and biota in the upper portion and the accumulation of weathering, leaching, and decay products below. The *water table* marks the transition from the vadose zone to the deeper *groundwater zone*, where the pore space is filled by water. Like the vadose zone, the groundwater zone is a chemically reactive reservoir. The lower limit of the Critical Zone generally corresponds to the base of the groundwater zone, a diffuse boundary of variable depth extending a kilometer or more below the surface. The volume of water in the upper kilometer of the continental crust is an order of magnitude larger than the combined volume of water in all rivers and lakes.[1] The Critical Zone is perhaps the most heterogeneous and complex region of the Earth and the only region of the solid Earth readily accessible to direct observation.

[1]NRC, *Opportunities in the Hydrological Sciences*. National Academy Press, Washington, D.C., 348 pp., 1991.

Global Climate Change and the Terrestrial Carbon Cycle

A significant amount of carbon is stored in soils and sedimentary rocks, thus the Critical Zone plays a key role in the global atmospheric CO_2 balance.[1] Soil constitutes the third largest carbon reservoir, and work with carbon-14 tracers reveals that the distribution of soil organic matter strongly influences the means and rate of carbon uptake and release. The exchange of carbon among atmosphere, ocean, and terrestrial reservoirs is also affected by human land-use practices and land-use histories (e.g., agriculture,

[1]*A U.S. Carbon Cycle Science Plan*. A Report of the Carbon and Climate Working Group, J.L. Sarmiento and S.C. Wofsy, co-chairs, U.S. Global Change Research Program, Washington, D.C., 78 pp., 1999 (http://www.carboncyclescience.gov/planning.html# plan).

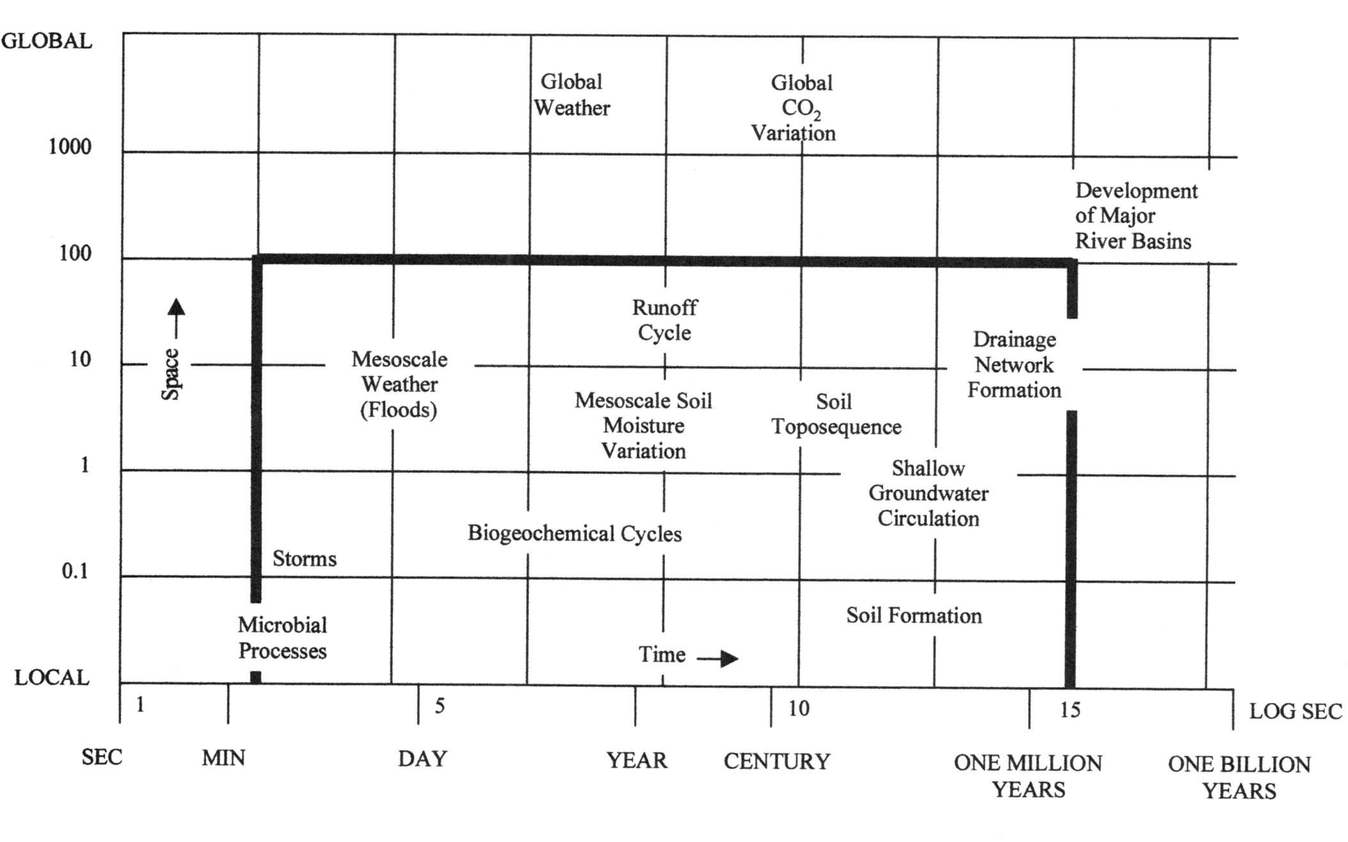

FIGURE 2.2 Spatial and temporal scales of Earth surficial processes, with those occurring in the Critical Zone enclosed in a solid frame. Modified from *Opportunities in the Hydrological Sciences*, National Academy Press, Washington, D.C., 348 pp., 1991. SOURCE: G. Sposito, University of California, Berkeley.

reforestation), as well as by silicate weathering. However, the temporal and spatial variability of carbon sources and sinks is not well documented, especially on longer time scales. New evidence regarding the influence of microscopic and macroscopic organisms on the weathering of silicate minerals and the transfer of soluble carbonates to the substratum may provide constraints on weathering reactions and global climate change.

The Interactions of Life, Water, and Minerals

Organic molecules and microorganisms strongly affect the kinetics of important geological and pedological processes. However, these materials studies are only beginning to address the quantitative significance of microbial interactions in mineral weathering, soil formation, and the geochemical cycling of metals, nutrients, and other elements or isotopes (Figure 2.3). For example, it is clear that organic compounds influence the burial of reduced carbon and help drive the decay or retention of toxins in natural soils and sediments, although mechanisms for the preservation of carbon are still being debated. The interactions of water with minerals and other materials in soils, sediments, and rocks are critical in the dispersal and concentration of chemical species, the migration of contaminants, and the accumulation of natural resources. The architecture of porous media influences transmission dynamics, sorption-desorption kinetics, and chemical fates. Models of physical transport in heterogeneous media are being developed to understand chemical and biological reactions, nutrient cycling, and the fates of contaminants. Such models are being constrained by imaging and analytical measurements developed over the past few years and will provide new insight into the physical, biological, and chemical influences on water quality and availability.

The Land-Ocean Interface

New technology is yielding vastly improved insights to the nature and dynamics of coastal sedimentary environments.[2] For example, scanning airborne lasers are measuring seafloor bathymetry and topography of coastal areas with unprecedented accuracy and spatial coverage allowing assessment and understanding of coastal change in ways that were not possible only a

---

[2]*Coastal Sedimentary Geology Research: A Critical National and Global Priority*, results of a workshop held in Honolulu, Hawaii, November 9-12, 1999.

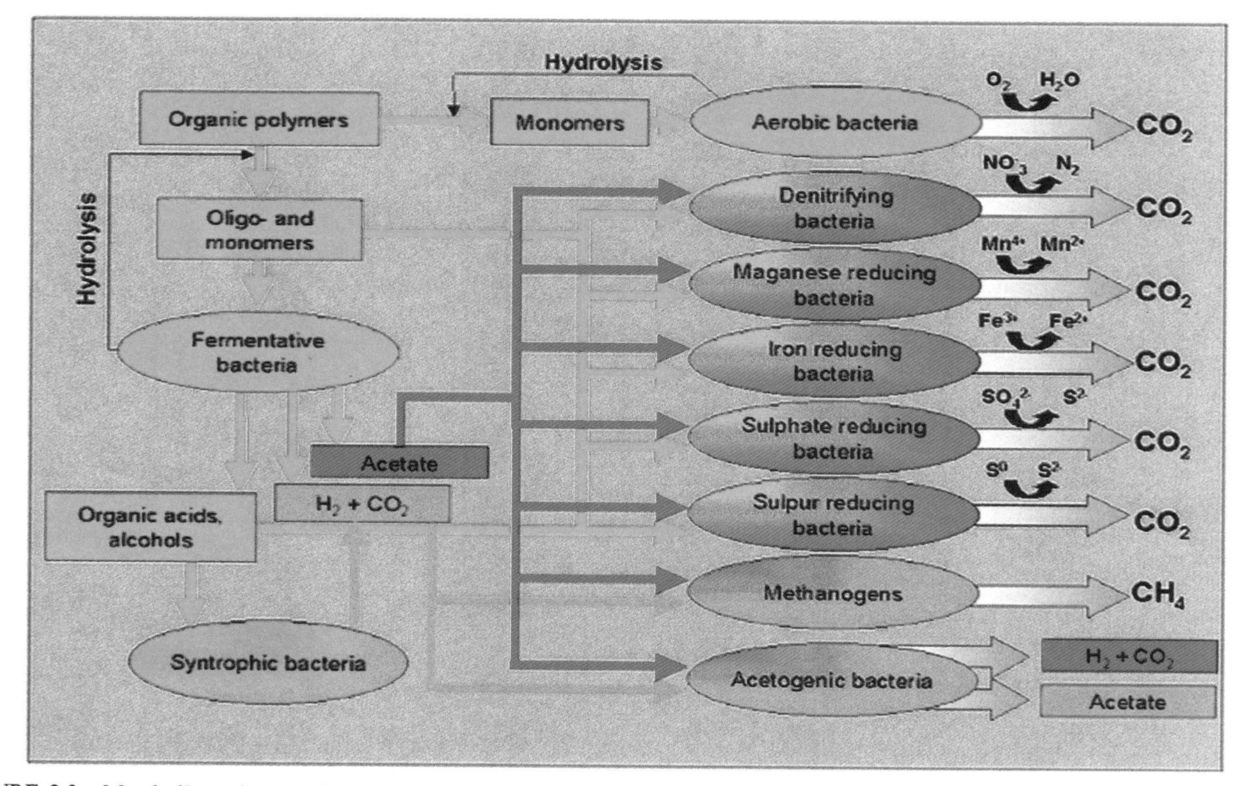

FIGURE 2.3 Metabolic pathways affecting the degradation of organic carbon. In nature, pathways for organic matter degradation are interlinked, with the final step depending on available oxidants. Oxygen is the most efficient reagent for respiration, but it can quickly be depleted in wet environments. In the absence of oxygen, bacteria will employ nitrate, metal oxides, sulfate, or sulfur to convert organic compounds into CO_2. Additional pathways include the conversion of complex compounds into simple substrates, fermentation, and the cycling of carbon dioxide, hydrogen, methane, and acetate by Archea. SOURCE: K. Pedersen, Göteborg University, http://pc61.gmm.gu.se/gmm/groups/pedersen/basic_research_at_the_dbl.htm.

few years before. Using lasers and spectral imagers, it is now possible to map seafloor habitats and to determine whether the substrate is fixed or movable, living or fossil. These data also provide a better basis for integrated three-dimensional models of coastal ocean processes (e.g., tides, waves, currents). New models that integrate land-related processes (e.g., groundwater flow, sediment flux and storage, morphology of drainage systems and watersheds) are being used to understand nutrient loading in the coastal zone, as well as the ultimate fate of waste materials and pollutants in the ocean.

Tectonics, Climate, and Weathering

Topography, surface hydrology, sedimentation, and climate are intimately related. For example, erosion of the Tibetan-Himalayan plateau is related to precipitation and glacier development. Climate and hydrology interact to modify the surface expression of tectonic events, which in turn influence the rate and scale of fluvial responses. Topography exerts a major influence on erosion and the subsequent deposition of sedimentary rocks, as well as the formation and landscape diversity of soils. Similarly, removal of materials by surface processes influences tectonic uplift rates. Coupled geologic-hydrologic-climate studies promise a greater understanding of denudation rates, weathering processes, and the survivorship of mountain ranges.

Earth History

The history of the planet is recorded in soils, sediments, ice, water, and rocks. The geological record of environmental variations during the last several hundred thousand years provides the context for understanding the current climate system and its potential for future change. Important insights can also be gleaned from the behavior of the Critical Zone over even greater spans of geologic time: topographic relief, length of day, solar influx, and composition of the ocean and atmosphere have all varied significantly in the past. A good example, based on both careful geological field work and geochemical and isotopic observations, is the recent suggestion that the Earth went through a series of global glacial events ("snowball Earth") about 750 million to 580 million years ago. The implications of such severe climatic conditions and other extreme events (e.g., extensive volcanism, meteorite impacts) for the evolution and maintenance of life on Earth are the subject of ongoing debate, stimulating new efforts to characterize extreme environmental conditions of the past.

New Tools and Observations

Many science disciplines—hydrology, geomorphology, biology, ecology, soil science, sedimentology, materials research, and geochemistry—are bringing new and powerful research tools to bear on the study of the Critical Zone as an integrated system of interacting components and processes. It is now possible to study the Critical Zone over a much greater range of length and time because of the wealth of data from satellites and aircraft that provide global information on scales from seconds to decades; advances in geochronology that extend the detailed record of near-surface environments to millions of years; imaging methods (e.g., electron and atomic force microscopes) and spectroscopic tools that probe soil materials to the atomic scale; and new information technologies, which permit the manipulation of large data sets and a variety of numerical simulations, from ab initio models of atomic and molecular interactions to global ocean and atmospheric circulation and mountain belt evolution. These technological advances have set up opportunities for novel cross-disciplinary activities.

- Synchrotron-based X-ray spectroscopy can be applied in tandem with computational geochemistry to elucidate the molecular-scale mechanisms of key aqueous geochemical reactions related to mineral weathering, contaminant sorption and desorption, and nutrient cycling.
- High-resolution electromagnetic and acoustic imaging of hydrological systems can be combined with molecular-scale mechanisms of aqueous geochemical reactions for accurate representations of transport in reactive systems. Such representations are essential for addressing silicate mineral weathering—a source of nutrients to the biosphere and a major control on the long-term CO_2 budget—and contaminant transport.
- The techniques of isotope geochemistry and molecular biology reveal the pathways involved in biogeochemical cycles and the formation of secondary minerals in weathering environments.[3] New geochemical and stratigraphic tools and techniques for dating individual minerals are furnishing insights into the behavior of the Earth's surface with increasing temporal resolution.
- Remote-sensing data, digital elevation models, and special dating techniques such as cosmogenic nuclide exposure ages can be used to validate a new generation of geomorphic transport models.[4] These models will allow a

---

[3] *Research Opportunities in Low-Temperature and Environmental Geochemistry*, results of a workshop held in Boston, Massachusetts, June 5, 1999.

[4] *A Vision for Geomorphology and Quaternary Science Beyond 2000*, Results of a workshop held in Minneapolis, Minnesota, February 6-7, 2000.

more quantitative exploration of the dynamic interrelationships between tectonics, climate, soil diversity, and landscapes.

- Microprobe, X-ray diffraction, scanning and transmission electron microscopy, X-ray tomography, and microchemical probes can be used to map the architecture of soils and to investigate material properties at the atomic-scale resolution needed for understanding sorption-desorption kinetics and other equilibrium processes.[5]
- Ground-penetrating radar and three-dimensional seismic imaging of sedimentary deposits permit modeling and prediction of physical properties of heterogeneous sediments in three dimensions.[6] Chemostratigraphic techniques can be used to correlate sediments among sedimentary basins, particularly between onshore and offshore basins.
- In situ and aircraft sensors for measuring circulation patterns and mapping the bathymetry and substrate of the near-shore environment, combined with analysis of geochemical and sedimentological components and fluxes, can be used to quantify the variability of the geological, biological, and atmospheric components of coastal ecosystems.

Need for Coordinated Field Work and Integrated Modeling

The integration of disciplinary research is the key to future progress in the science of the Critical Zone. This theme permeates the discussion of many other aspects of Earth science in this report, but, in the case of the Critical Zone, it presents some special challenges, in part because of the sheer number of the disciplines and the diversity of their approaches, but more profoundly because of the spatial scales intrinsic to the scientific issues. Although Critical Zone processes often involve the global aspects of atmospheric and oceanic transport, many of the most intense interactions occur in relatively localized regions of the solid Earth—for example, over dimensions less than the characteristic horizontal variations in topography and near-surface geology (tens to hundreds of kilometers) or the thickness of the zone itself (about a kilometer). Indeed, much of the science to be done will require the in situ study of microscopic processes that are subject to numerous contingencies—physical, chemical, and biological—which vary from one surface environment to the next. Not surprisingly, disciplinary integration has proceeded more

---

[5]*Opportunities in Basic Soil Science Research*, G. Sposito and R.J. Reginato, eds., Soil Science Society of America, Madison, Wisconsin, 129 pp., 1992.

[6]*Sedimentary Systems in Space and Time: High Priority NSF Research Initiatives in Sedimentary Geology*, results of a workshop held in Boulder, Colorado, March 27-29, 1999.

rapidly in the study of global-scale geosystems, which manifest a more obvious set of unifying concepts, diagnostic behaviors, and simple symmetries—for example, the dipolar magnetic field (core dynamo), the shifting mosaic of plate tectonics (mantle convection), or the largely zonal structure of global circulation (climate system).

New mechanisms are thus needed to encourage multidisciplinary collaborations on Critical Zone problems, especially on local and regional scales. Modeling activities that employ conceptual and numerical tools to integrate different types of data are clearly important. However, the primary deficiency at this stage of the science is the difficulty in mounting field work to collect measurements that are sufficiently localized and simultaneous as well as dense and comprehensive enough to constrain process-based models of Critical Zone behaviors. Real progress will require some way to coordinate the field investigations of hydrologists, pedologists, geochemists, geobiologists, mineralogists, and other geoscientists in localized regions, often for extended periods of observation, and to encourage the integration of these data with controlled laboratory measurements and system-level models.

One mechanism for encouraging this type of problem-focused, multidisciplinary field work is through the establishment of "natural laboratories" in which detailed, long-term observations can be made using a variety of disciplinary tools. As discussed in Chapter 3, such a program would also provide new opportunities for scientific advancement in many other areas of Earth science.

GEOBIOLOGY

Life is inextricably linked to the Earth, so it is not surprising that some of the most challenging problems of geology and biology are intimately interwoven. The synthesis of these two sciences is geobiology, which addresses the interactions of biologic and geological processes, the evolution of life on Earth, and the factors that have shaped the current and past biospheres. Important issues include the following:

- origin of prebiotic molecules and life, and its early evolution;
- emergence and divergence of metabolisms and morphologies;
- effects of organisms on the physical and chemical characteristics of Earth and its fluid envelopes;
- nature of ecosystems and their response to environmental perturbations of many types; and
- the rules that govern biodiversity dynamics, including selectivity in

extinction and recovery, interactions among groups, and the effects of geographic scale.[7]

Much can be learned from studies of present-day systems, but access to the geological record also permits these issues to be explored over longer time scales and over a wider range of Earth system states. Major scientific breakthroughs are inevitable because of the enormous growth in the power of new techniques and of conceptual advances in the contributing biological, geological, and geochemical disciplines. Geobiological research results bear directly on a wide array of important scientific and societal issues, including the nature of the Critical Zone, the stability and resilience of ecosystems under stress, and the origin and evolution of life itself.

Geobiological processes operate over a wide spectrum of temporal and spatial scales—from small-scale, rapid exhalation of O_2 by cyanobacteria, to cycling of carbon through communities of marine phytoplankton and rain forests, to restructuring of regional and global ecosystems in the wake of major extinction events, to shifts in reef composition in response to slowly changing rates of seafloor spreading (Figure 2.4). Understanding the complex interactions that occur across such a broad space-time spectrum is challenging and requires an integrated approach that treats ecosystems as components of larger geosystems.

Recent Advances

Over the past decade, Earth scientists have gained much better appreciation of the impact of living organisms on their surroundings, which is forcing a critical reexamination of the relative importance of the various factors that control Earth processes.

Climate Studies

Biominerals provide a sink for atmospheric CO_2, and their formation is directly linked to the supply of calcium and magnesium from chemical weathering of silicate and carbonate minerals. Organic by-products may, in turn, increase dissolution rates of silicate minerals by factors ranging from 2 to more than 100. Such biological-climate feedback mechanisms (including

---

[7]See also *Dynamic History of the Earth-Life System: A Report to the National Science Foundation on Research Directions in Paleontology*, results of a workshop held in Washington, D.C., March 6-9, 1999.

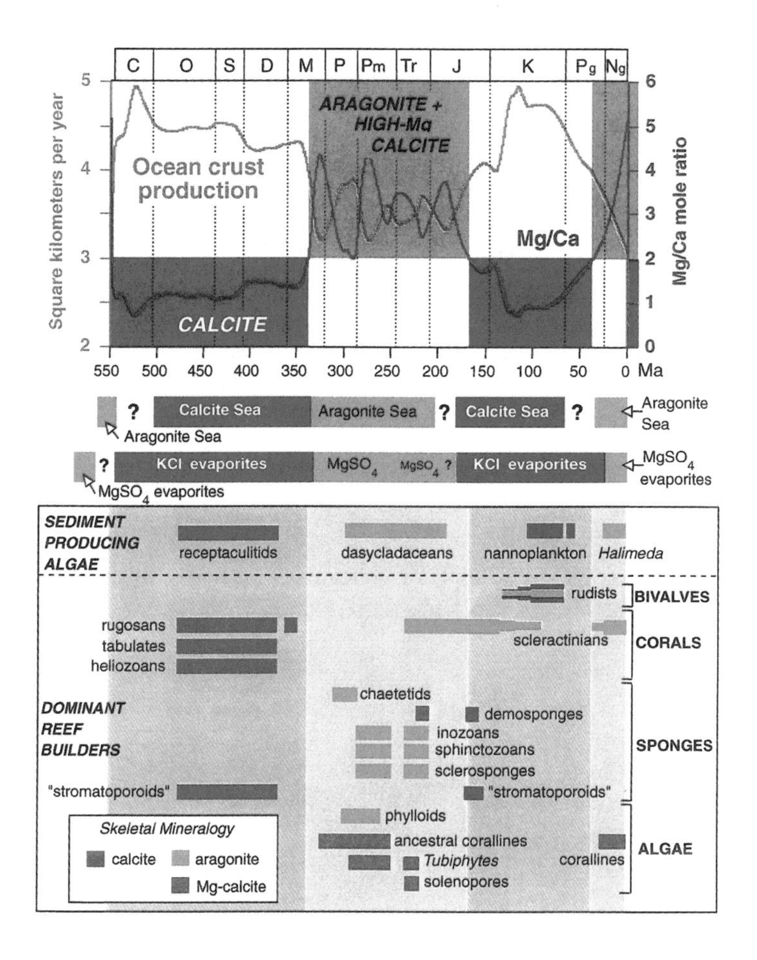

FIGURE 2.4 Temporal correlation between seafloor spreading rate (*top*) and skeletal mineralogies of sediment-producing algae and dominant reef builders through the Phanerozoic (*bottom*). The seafloor spreading rate and mineralogies of dominant hyper-calcifying marine taxa are thought to be linked through the oceanic magnesium to calcium ratio, here inferred from the ocean crust production rate. The magnesium-calcium ratio is also thought to have played a role in observed patterns of evaporite deposit mineralogies (*middle*). Magnesium-calcium ratios less than 2 yield "calcitic" seas whereas ratios greater than 2 yield "aragonitic" seas. SOURCE: Modified from S.M. Stanley and L.A. Hardie, Hypercalcification: Paleontology links plate tectonics and geochemistry to sedimentology, GSA Today, v. 9(2), p. 1-7, 1999. Modified with permission of the publisher, the Geological Society of America, Boulder, Colorado USA. Copyright ©1999 Geological Society of America.

the albedo of land vegetation, burial of organic carbon, production of biogenic compounds such as dimethyl sulfide that affect cloud formation, and linkages between oceanic biological productivity and iron and phosphorus supply) are becoming important parameters in global change models. Pollen data from cores, corroborated by isotopic and chemical evidence from reef corals, equatorial glaciers, and groundwater, provided the first evidence of tropical cooling during the Last Glacial Maximum, causing reconsideration of heat transport in global circulation models (GCMs). Paleobotanical evidence from the western United States in the late Paleocene has significantly refined the treatment of orographic effects in GCMs during rapid warming events, and evidence of extensive warm climate vertebrates and plants near the Cretaceous poles provides important constraints for modeling times of extreme warmth.

Biological Controls on Earth Chemistry

The pathways and extent to which organisms fundamentally control the distribution of elements (and isotopes) in the Earth's crust are now beginning to be understood. Most of the iron, sulfur, phosphorus, carbon, and nitrogen in soils and sediments pass through biological repositories, and there is growing evidence that metabolic processes have detectable effects on other major (e.g., Ca, Si) and minor (e.g., As, Mo, U) elements in seawater. Results of such studies provide the mechanistic basis for modeling biogeochemical cycles and for environmental studies of the mobility and fate of toxic compounds. New insights into the roles of microbial metabolism, organic chelators, and sediment-irrigating and -advecting plants and animals in processes such as mineral dissolution, mineral precipitation, soil formation, and sedimentary diagenesis are opening the way for more detailed understanding of Critical Zone processes both modern and ancient.

Molecular Geobiology

New data from molecular biology, developmental biology, and paleontology have invigorated studies of the origins, evolution, and ecology of major biological groups. Important advances include the construction of evolutionary trees of all living things, based primarily on genetic similarities of living taxa, and the placement of the major branching events of these trees into their correct geological context (Figure 2.5). Improvements in instrumentation and preparative techniques have made it possible to identify and characterize the isotopic composition of biomarker molecules in sedimentary rocks, providing evidence for the operation of metabolic pathways important in biogeochemical

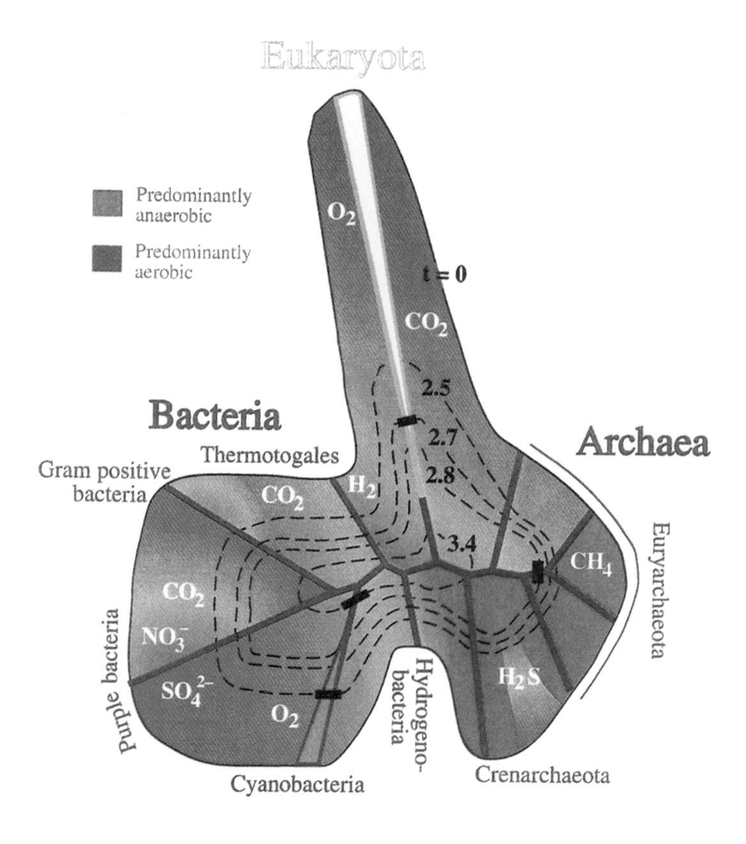

FIGURE 2.5 Evolutionary tree showing the relationship between the major branches of life. Some key metabolic by-products (e.g., CO_2) of the major groups of organisms are labeled in white. Orange and green colors represent aerobic versus anaerobic environments. Contours, representing time in billions of years, are schematic and are anchored only by a few minimum dates (black boxes) gleaned from the geologic record. Correlating genomic events with the geological record remains a significant challenge. Where the geologic record indicates considerable diversity within a group, the line depicting the lineage has been widened (yellow and blue wedges). SOURCE: Reprinted with permission from J.F. Banfield and C.R. Marshall, Genomics and the geosciences, *Science*, v. 287, p. 605-614, 2000. Copyright 2000 American Association for the Advancement of Science.

cycles through time. Key discoveries include molecular confirmation that cyanobacteria were already producing major amounts of oxygen 2.5 billion years ago and that organisms with structurally complex eukaryotic cells were possibly present at least 2.7 billion years ago, 500 million to 600 million years earlier than the oldest known eukaryotic fossils.

The discovery that most living animals share the same core set of developmental genes has begun to make it possible to understand the genetic basis for the differences between animals. Integration of these genetic data with new morphological and geochronologic data from the fossil record allows unraveling of the evolutionary events that led to novel morphologies, from the earliest multicellular animals to the origin of flowers and the vertebrate skull. This offers the opportunity for understanding key events, such as the Cambrian explosion of multicellular animals (Figure 2.6) and the mid-Paleozoic invasion of land. Geological and molecular data can also be used to calibrate molecular clocks based on differences between the DNA of living species; these clocks can then be applied to lineages otherwise lacking a rich fossil record.

Discoveries of entirely new microbial organisms in ecologically and geologically extreme environments (extremophiles) have greatly broadened the concept of the versatility of life; microorganisms have been found kilometers deep in Antarctic ice, in active vents along midocean ridges, and in fluids with a pH of zero. Subduction zones likely contain the Earth's deepest biota, sustained by energy from chemical redox reactions rather than solar radiation.

Studies of organic reactions on mineral surfaces have led to a new understanding of how key prebiotic compounds were formed early in Earth history. Molecules such as thioesters and acetic acid have been created under geochemical conditions relevant to the early Earth, and nucleotides and amino acids have been polymerized, providing possible clues to the origin of life.

Evolutionary Innovations

One of the most striking paleontological observations has been the uneven distribution of evolutionary innovations in time and space. Pulses of innovation following extinction events, such as the early Cenozoic radiation of mammals following the extinction of dinosaurs and other Mesozoic dominants, are now recognized as a crucial component of the evolutionary dynamic and are the focus of a major new research direction linking paleobiologists, evolutionary biologists, and ecologists. Evolutionary innovation also has a strong spatial component, with major novelties appearing in disturbed environments, both on land and in the sea, and in tropical latitudes. Recoveries from extinction events and diversifications show significantly different trajectories in different regions that sum to lasting effects on the global biota (Figure 2.7).

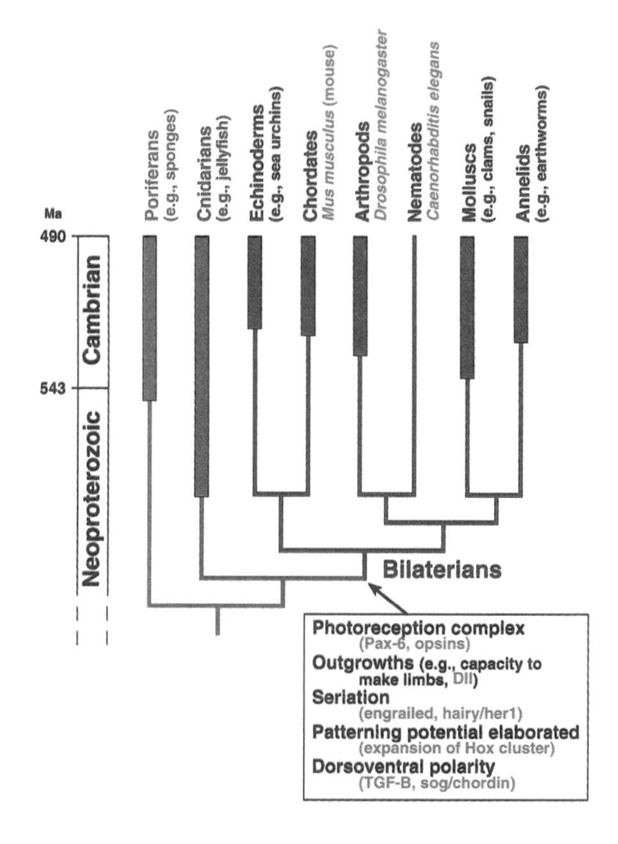

FIGURE 2.6 Phylogeny of selected animal phyla, showing the stratigraphic extent of their fossil records (body fossils only). Inferred character states of the last common ancestor of the bilaterian animals are listed in the box. The distribution of key developmental genes, along with comparative anatomical data, can be used to infer the developmental potential and soft tissues present in ancestors of these phyla, even though no fossils of these ancestors exist. Most of the developmental data come from a few living "model" systems (species names in red). Some of the genes identified from these and other species that play a role in developmental capacity or tissue (indicated in the box) are also shown in red. There is great uncertainty in the time of divergence of the animal phyla: the fossil record is consistent with divergence shortly before the beginning of the Cambrian, whereas molecular clocks suggest divergence perhaps as much as 1000 million years ago. SOURCE: Modified with permission from A.H. Knoll and S.B. Carroll, Early animal evolution: Emerging views from comparative biology and geology, *Science*, v. 284, p. 2129-2137, 1999. Copyright 1999 American Association for the Advancement of Science.

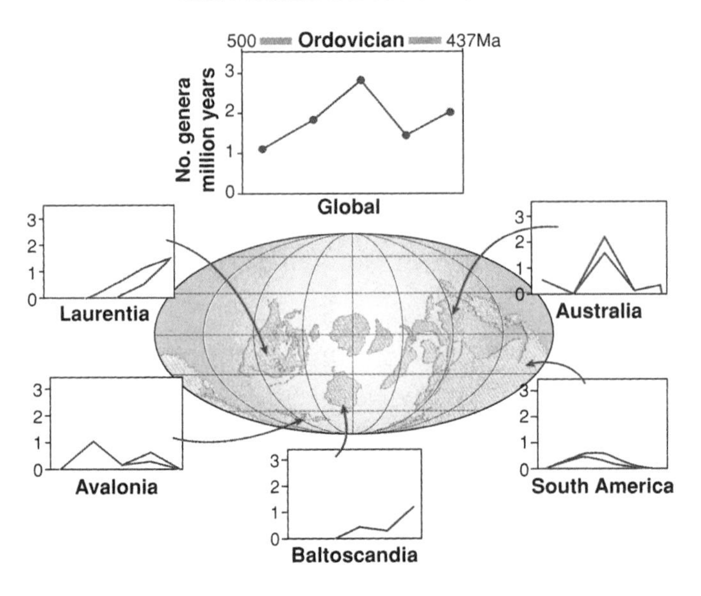

FIGURE 2.7 Patterns of diversification of Ordovician bivalve molluscs around the world. As with other marine organisms, the pattern varies by geographic location, indicating that the Ordovician radiation (the largest diversification of animal genera seen in the fossil record) was not a globally synchronous event. Analysis of the geological histories of these regions suggests that local tectonics played a role in the creation of niches required for diversification to occur. SOURCE: Reprinted with permission from D.H. Erwin, Evolution —after the end—recovery from extinction, *Science*, v. 279, p. 1324-1325, 1998. Based on data from A.I. Miller, Comparative diversification dynamics among paleocontinents during the Ordovician radiation, *Geobios*, MS 20, p. 397-406, 1997, and A.I. Miller, personal communication, and map from C.R. Scotese and W.S. McKerrow, in *Palaeozoic Paleogeography and Biogeography*, W.S. McKerrow and C.R. Scotese, eds. (Geological Society of London Memior 12, p. 1-21, 1990). Copyright 1997 American Association for the Advancement of Science.

Environmental and Ecological Dynamics

Paleobiological analysis of the geologic record has provided insights into the dynamics of ecological communities that could never be discovered through annual or decadal observations of modern systems or by theory alone. For example, data on North American land plants, insects, vertebrates, and marine invertebrates show that many species shifted their geographic ranges independently, rather than moving as cohesive communities, in response to climate changes since the Last Glacial Maximum. This observation has important implications for the response of natural and agricultural systems to anthropogenic climate change. Comparative analyses of other events in the geological record are revealing the complexity of biotic response. For example, the Late Paleocene Thermal Maximum, possibly enhanced by a massive release of methane hydrates at ocean margins, produced a surprisingly mild extinction, and in some environments even promoted diversification, outside of the deep sea.

New Tools and Capabilities

New and improved biological, chemical, geological, and paleontological methods hold enormous potential for geobiological research. These capabilities allow investigations that would have been virtually inconceivable in past decades.

- Technical advances in molecular approaches such as genomics (the mapping and sequencing of genomes and analysis of gene and genome function), proteinomics (characterizing the structure-function relationship of proteins), and developmental biology provide a means of (1) assessing the diversity of organisms and of biological metabolisms in modern and ancient environments; (2) tracing the role of gene transfer between species, which is relevant to the evolution of life; (3) estimating phylogenetic relationships within and among major microbial, plant and animal groups; and (4) unraveling the mechanistic basis of morphological innovation.
- Combinations of powerful solid, surface, solution, and organic analytical techniques with conventional and molecular biological approaches can now be used to (1) tackle the dynamics of trace and toxic element cycling between solutions, solids, and biosphere reservoirs; (2) unravel the mechanisms of biomineralization and the pathways of microbial degradation of organics; (3) study the ecology of macroorganisms (through analyses of bones, teeth, and shells); and (4) develop kinetic models for biogeochemical processes.
- New molecular and isotopic techniques, applied to the remains of organisms and in analysis of fossil morphologies, are being developed for tracking levels of atmospheric CO_2 and O_2 and differentiating oceanic water masses and their movements over geologic time. The combination of these methods with expanded computer models and a better understanding of ocean and atmospheric circulation will greatly advance knowledge of the factors that affect climate and the geochemical and biological cycling of key elements and molecules in the environment.
- Recent technical improvements have made it possible to measure radioactive and stable isotopic differences with much higher precision and

on much smaller samples than was previously possible. New methods of age interpolation (e.g., based on periodic orbital forcing functions), data integration, and quantification of gap effects and diachrony in geochronological datums are leading to significant refinements of the geological time scale. The increased temporal acuity offers the opportunity to test the plausibility of mechanisms proposed for major events seen in the rock and fossil records, such as the end-Permian mass extinction.

- Comprehensive studies of past and present environments, combined with numerical simulations and "meta-analyses" of diverse data sets, are dramatically improving our insight into the quality of the geological record. New methods for estimating the completeness of sedimentary successions and evaluating hiatuses (Figure 2.8) are leading to improvements in the sampling and analysis of the rock and fossil records, and field and laboratory studies are leading to a greater understanding of the processes of selective modification and preservation of the geologic record.

In addition to new tools, advances in geobiology depend on new field studies of regions and key stratigraphic intervals, phylogenetic and systematic analyses of key groups, and reevaluation and organization of existing data archived in museums, written records, and electronic media. These investigations yield vital new information on the geography, environmental and evolutionary context, age, and composition of biotas and on the deployment of evolutionary lineages in time and space.

Science Goals and Challenges

Recent advances in relevant technologies and capabilities have set the stage for significant progress in understanding the role of biological processes in geosystems, the evolution of life on Earth, and the factors that have shaped the current and past biospheres. Fundamental new knowledge at the interface of the biological and Earth sciences is anticipated in the following broad areas:

- the interrelationship among organisms, biological processes, and the fundamental physical and chemical properties of the Critical Zone;
- the extent to which geological processes, such as weathering and mineral precipitation, are mediated by biological processes, and the mechanisms by which this mediation occurs;
- the ways in which biogeochemical cycles operate and are controlled or modulated by physical, chemical, and biological processes;
- the role of physical factors, including rare events, in structuring and changing the composition and organization of biological communities;

- the relative importance of intrinsic and extrinsic factors in controlling the nature and rates of biological innovation;
- the rates and selectivities in biological extinction and recovery, and how these scale to duration, magnitude, and geographic extent of changes in the physical and biological environment; and
- the causes of environmental change, including the ways in which biological processes contribute to and can be used to remediate environmental damage.

EARTH AND PLANETARY MATERIALS

Research on Earth and planetary materials has undergone significant growth over the past two decades, building on major advances in the well-established domains of geochemistry, mineralogy, petrology, and soil science, as well as in the newer fields of geomicrobiology and biomineralogy, mineral and rock physics, and nanophase and planetary materials studies. This research is also being stimulated by new collaborations of geoscientists with chemists, physicists, molecular biologists, and materials scientists. Together, these disciplines have laid the groundwork for the ambitious objective of attempting to understand, from the atomic level up, the most significant processes that determine the current state and geological evolution of planets in general and Earth in particular.

Recent Advances

Earth and planetary materials research is based on an atomistic approach—establishing properties at the molecular level in order to understand materials and processes at much larger scales. It has involved the development of major new research tools. For example, mineral physicists have been at the forefront of developing synchrotron beamlines for micro-diffraction and spectroscopy, experimental tools for studying materials at ultrahigh pressures and temperatures, resonance techniques for superprecise measurements of elastic properties, and quantum mechanical methods for modeling complex minerals. Characterizing Earth materials with such methods is essential for quantifying the underlying processes and driving forces at play.

These capabilities also make it possible, for the first time, to study naturally occurring nanophases and mineral surfaces in great detail, for instance, tracking low concentrations of elements in soils and related biological materials. Understanding how both naturally occurring and human-introduced

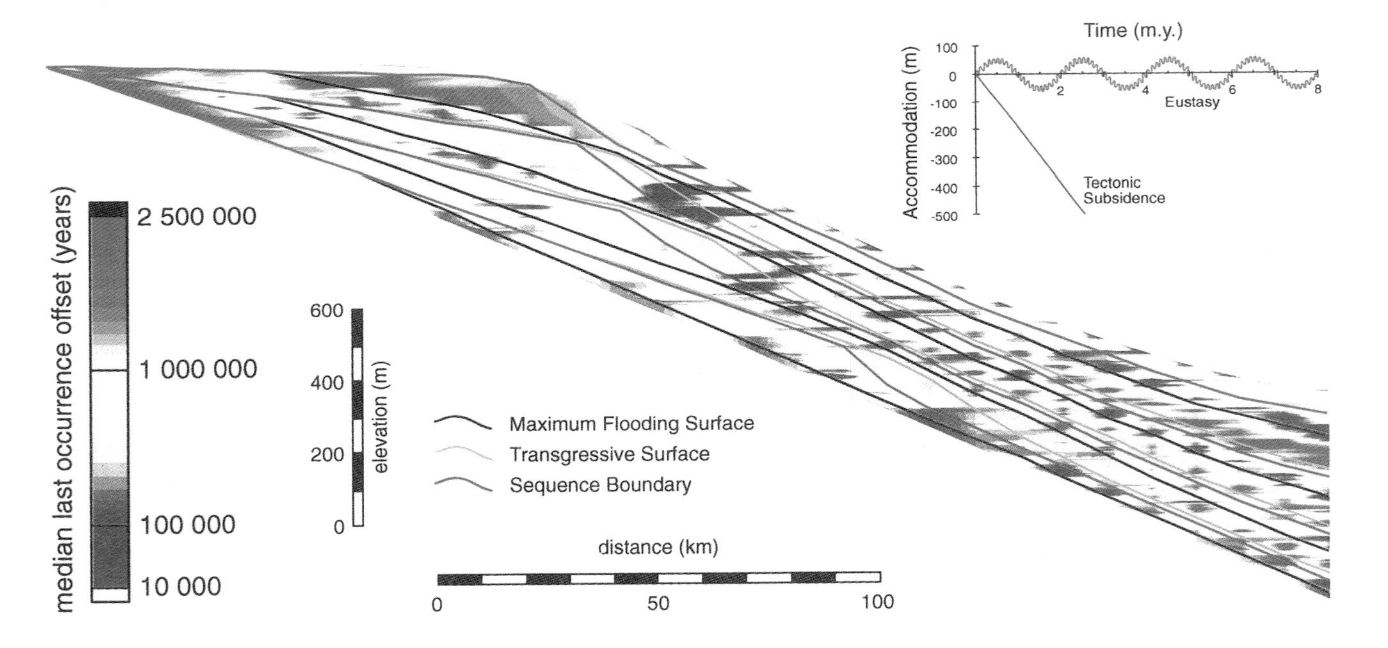

FIGURE 2.8. Two-dimensional synthetic section across a sedimentary basin from onshore (*left*) to offshore (*right*). The model used to produce the synthetic section includes four eustatic sea-level cycles, producing four depositional sequences, superimposed on a constant rate of tectonic subsidence (upper right). Maximum flooding surfaces, transgressive surfaces, and sequence boundaries represent time lines through the section. While the sequences were being deposited, species with different ecological (depth) preferences were allowed to evolve and to become preserved in the forming rock record. The colors represent differences in each species' true time of extinction and the time of extinction observed in the synthetic section: high fidelity of the fossil record is shown in blue, and low fidelity is shown in red. The fossil record is poor onshore and offshore above the transgressive surfaces. Such models are useful for understanding large-scale biases in the sedimentary rock and fossil records. SOURCE: Reprinted with permission from S.M. Holland and M.E. Patzkowsky, Models for simulating the fossil record, *Geology*, v. 27, p. 491-494, 1999. Copyright 1999 Geological Society of American.

elements are released by soil, biologically cycled, and moved through the food chain is important to ensuring food security and safety while maintaining or improving environmental quality.

It is now possible to reproduce experimentally the high pressures and temperatures existing throughout the Earth and to perform accurate measurements of material properties at these conditions. This is significant because seismology is revealing the structure of the Earth's deep interior with rapidly improving resolution. By comparing seismological observations with laboratory measurements and theoretical analyses (Figure 2.9), it is possible to determine the state and composition of the planetary interior as well as the processes by which it evolves.

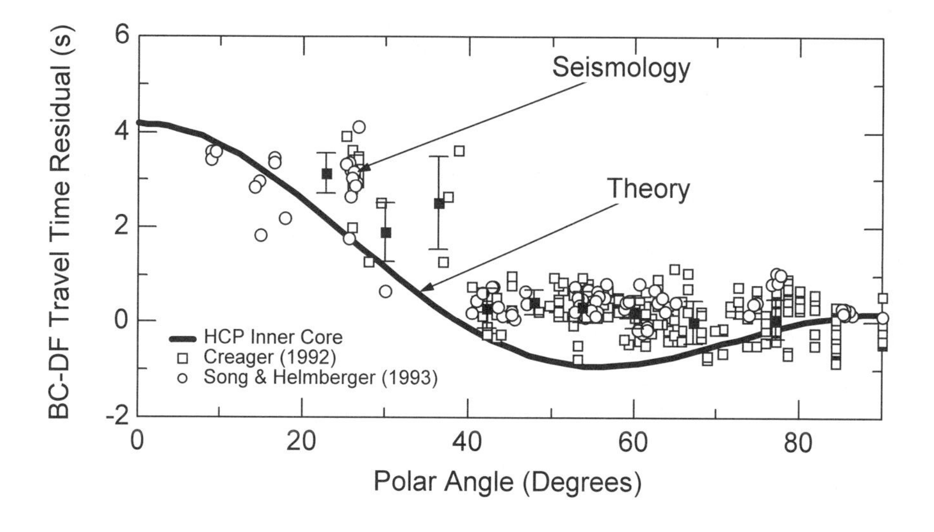

FIGURE 2.9 Quantum mechanics at the Earth's center. Theoretical calculations (curve) of the elastic properties of hexagonal close-packed (HCP) iron at conditions of the inner core reproduce seismologically observed variations of travel times at depth (points). The observations represent the time required for seismic waves to travel through the inner core, showing that travel times (hence wave velocities) vary systematically with propagation direction, from polar (along the Earth's rotation axis: polar angle = 0°) to equatorial (polar angle = 90°). The theory is from "first principles," meaning that the quantum mechanical calculations are free of any experimental input. SOURCE: Modified from *Microscopic to Macroscopic: Opportunities in Mineral and Rock Physics and Chemistry*, results of a workshop held in Scottsdale, Arizona, May 28-30, 1999. Reprinted with permission from L. Stixrude and R. E. Cohen, High-pressure elasticity of iron and anisotropy of earth's inner core, *Science*, v. 267, p. 1972-1975, 1995. Copyright 1995 American Association for the Advancement of Science.

Recent scientific developments are illustrated in the following examples:[8]

- Experimental discovery and initial characterization of the high-pressure metallic state of hydrogen, considered to be the predominant material making up the interiors of giant planets and stars—its existence was theoretically predicted 60 years earlier, but the new measurements reveal unexpected properties for metallic hydrogen and are leading to significantly improved models of the structure, dynamics, and evolution of giant planetary interiors. Understanding giant planets is required to decipher the origin of planetary systems (including our own) and is further motivated by the recent discovery of giant planets outside the solar system.
- The friction laws for rock, including its dependence on sliding velocities, have been determined experimentally for the first time, providing new insights into the physics of earthquake rupturing and unifying many observations of seismic phenomena that are relevant to assessing earthquake hazards. Moreover, rock has been found to exhibit highly nonlinear properties (e.g., in elastic wave propagation) among other reasons, because of the chemical and physical interactions of fluids with the formation and propagation of fractures at microscopic to tectonic scales. Nonlinearity means that small forces can have large effects.
- Newly discovered mineral phases that are formed at high pressures lock water, carbon dioxide, and other "volatile" molecules into their crystal structures. The Earth's interior is thus likely to contain far more water and other volatile species than the hydrosphere, fundamentally altering current views of how the oceans and atmosphere have evolved over geologic time.
- Novel techniques are revealing for the first time the physical structures and chemical properties of mineral surfaces (Figure 2.10). Experimental measurements and theoretical analyses are providing a molecular-scale understanding of the detailed interactions between fluid (within pores or along grain boundaries) and mineral phases, with far-reaching implications for disciplines ranging from volcanology to seismology and for applications ranging from resource extraction to environmental remediation. In particular, the details of fluid-mineral interactions determine the degree, rate, and paths with which surface and underground contaminants migrate and the means by which such effects can be mitigated.
- Land resources are composed of soils with active colloidal fractions dominated by organic-mineral nanophases whose surfaces control the chemical speciation and fate, mobility, bioavailability, reactivity, transport,

[8]See also *Microscopic to Macroscopic: Opportunities in Mineral and Rock Physics and Chemistry*, results of a workshop held in Scottsdale, Arizona, May 28-30, 1999.

and toxicity of these constituents. Such knowledge has important societal and environmental implications in the Critical Zone, for example, in the sustainability of agriculture and water resources.

Science Opportunities

The momentum associated with recent discoveries, improved facilities, and new collaborations between disciplines offers a glimpse into future possibilities. The coming decade will see the emergence of major opportunities for research on Earth and planetary materials.

Development of Intense Neutron Beams and Other Powerful New Probes of Material Properties

Neutrons are uniquely suited for documenting the molecular sites of hydrogen (including water and hydroxide) in minerals and fluids; the textures of rocks, soils, and other polycrystalline materials; the atomic configurations in fluids; and the thermodynamic properties of crystals. Dedicated synchrotron X-ray beamlines, fully instrumented for sophisticated in situ experiments, will lead to an entirely new class of experiments with hundredfold or greater increases in spatial, temporal, and spectral resolution. These technical developments will, with little doubt, lead to breakthroughs in applied as well as basic research, from monitoring fluid-mineral interactions to clarifying the evolution of planetary interiors.

Retrieval of Samples From Other Planets, Interplanetary Space, and Comets

Future space missions, already initiated or planned, are expected to provide samples of cometary, asteroidal, and solar materials that are representative of the substances from which the planets formed. Specimens may also be returned from Mars within 10-20 years. These samples will be small and are expected to contain a significant fraction of dust-size particles, putting a premium on the use of modern high-resolution analytical techniques. Mineral surfaces and interfaces will require special study, especially to document potential precursors of biological molecules and organic processes in extraterrestrial materials, which will be a high priority for identifying any traces of life.

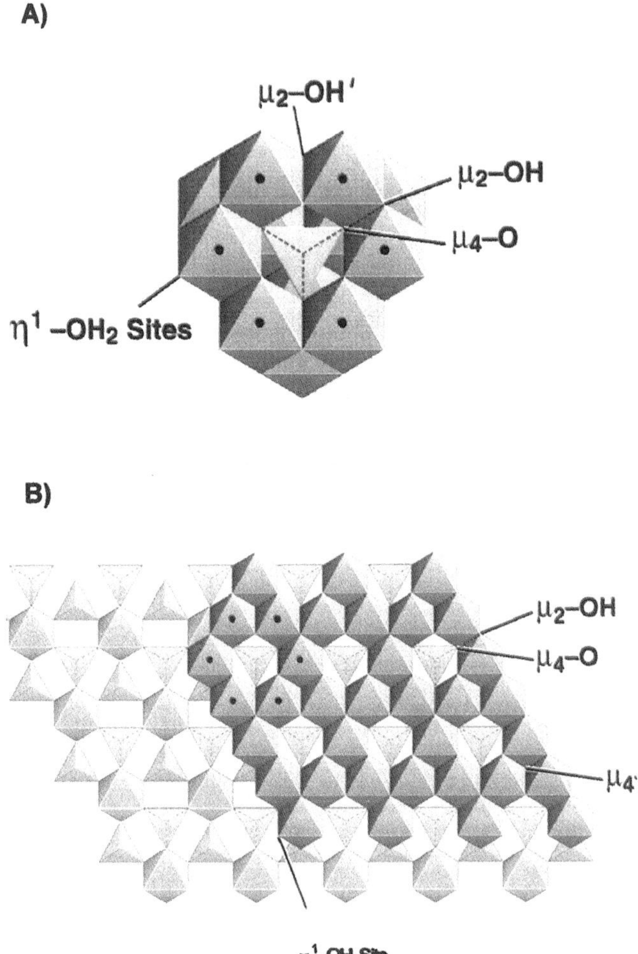

FIGURE 2.10 Dissolution of minerals. Experiments reveal the detailed processes controlling mineral-water interfaces, with the molecular arrangement of aluminum within the water (• in A) being found to closely match that at the surface of a γ-alumina crystal (• in B). Shaded octahedra and tetrahedra represent AlO_6 and AlO_4 molecular units, respectively. SOURCE: B.L. Phillips, W.H. Casey, and M. Karlsson, Bonding and reactivity at oxide mineral surfaces from model aqueous complexes, *Nature*, v. 404, p. 379-382, 2000. Reprinted by permission from *Nature* copyright 2000 Macmillan Magazines Ltd.

Application of Molecular Biology to Earth Materials

Microorganisms and nanocrystalline phases are now recognized as playing a key role in weathering and other fluid-soil-rock interactions within the Critical Zone. Geomicrobiology is just now gaining momentum and has already uncovered major surprises about how organisms influence mineralogical and geochemical processes in the natural environment. Molecular biological technologies offer powerful means for identifying both the species and the functionality of organisms in a wide variety of environments (see "Geobiology" above). Only a small fraction of the Earth's surface has been examined so far, suggesting that many significant discoveries remain to be made, including extreme habitats at depth in the crust.

Long-Term Observations in Natural Laboratories

Field observations are required to study material properties and processes on length and time scales inaccessible in the laboratory, as well as to assess which processes, among the many possibilities, are most significant in controlling material properties under natural conditions. In situ measurements of geological materials will be an important complement to laboratory experimentation for monitoring and remediating the environment, characterizing geological resources, and assessing geological hazards. Natural laboratories have enormous potential as facilities where data sets with appropriate detail and precision can be collected systematically over the long term under natural conditions. For example, the San Andreas Fault Observatory at Depth (SAFOD), proposed as part of the EarthScope initiative (Box 2.2), will use deep drilling to recover samples and collect in situ data on the fault zone properties that will permit a better interpretation of laboratory experiments on rock friction and failure.

These opportunities in Earth and planetary materials research involve technical advances, research efforts, or space missions that have already been initiated. There are probably many other unforeseen developments that will take place over the coming decade. It is new linkages between geoscientists and researchers in other disciplines, such as chemistry, physics, and molecular biology, that have led to many of the recent breakthroughs as well as several of these opportunities.

The overall objective of research on Earth and planetary materials is to obtain a full understanding of geological processes starting at the atomic scale. Modern instrumentation, along with quantum and statistical mechanical computations, are providing powerful new capabilities for achieving this objective. Although many materials remain to be characterized, the accomplishment of the past decades has been to identify many of the key materials and properties controlling the Earth's geological processes. A major intellectual challenge that remains is to understand how material properties and processes can be scaled from laboratory to planetary scales of distance and time.

THE CONTINENTS

Almost the entire span of Earth history is recorded in the continental crust. Rocks on the surface of the continents provide an archive reaching back at least 4 billion years, 20 times beyond the oldest oceanic crust and almost nine-tenths the age of the solar system. Thus the continents provide the only accessible long-term record of the Earth.

The continents have grown by magmatism, deposition of sediments, and crustal accretion and have been reshaped through tectonism, metamorphism, and fluid flux from the deep Earth. Properly interpreted, continental geology yields information about these processes and about the large-scale dynamics that control mantle convection, planetary differentiation, plate motions, and climate throughout Earth history. Today, many fundamental and controversial ideas about the continents await testing through the integration of field geology and geomorphology with new technologies in geochemistry, geophysics, and geodesy.

In recognition of the need for broadly based, multidisciplinary studies of the continents, EAR established the Continental Dynamics (CD) Program in 1982 with the objective of augmenting its highly successful core programs in tectonics, geophysics, and geochemistry (Appendix A). The CD program has flourished, sponsoring many regional investigations of the continental crust and upper mantle around the globe, encouraging the development of new tools and setting the stage for a new phase of investigation.

Recent Advances

Many of the most significant advances of the past decade have involved integration of new remote-sensing data and computational technologies with field and laboratory observations.

- The advent of geodetic techniques based on the Global Positioning System (GPS) and interferometric synthetic aperture radar (InSAR) satellites has made it possible to observe the motions of crustal fragments on time scales from seconds to decades and to record relative positions accurate to a few millimeters over baselines of thousands of kilometers. Through the combination of space geodesy and ground-based geological mapping, modern plate boundaries can now be dissected with an unprecedented level of precision (Figure 2.11), paving the way for advances in earthquake physics and crustal rheology.

- Construction of precise topographic data sets and digital elevation models has reinvigorated the study of active tectonics and surface processes though the analysis of landforms and river drainages. The quantification of topographic relief has been joined with process-oriented field work and simulations based in dynamical systems theory to explore the strong interactions between tectonism, erosion, and climate.
- Isotope geochronologists have dated rocks as old as 4 billion years with an uncertainty of less than 1%, making it possible to measure the rates at which major, sometimes catastrophic, events occurred in the distant past. These constraints have provided a much more precise temporal framework for investigating the field relationships among geologic, climatic, and biologic phenomena (e.g., volcanic eruptions, mass extinctions, and major glacial events).
- Seismic tomography has provided three-dimensional images of the continents that show the strong correlation between surface geology and mantle structure. The ancient cratons are underlain by deep "keels" extending to depths of several hundred kilometers, raising important questions about the composition and evolution of the subcontinental mantle that can be addressed by geochemical observations. Measuring the directional dependence of seismic wave propagation (seismic anisotropy) and comparing this with laboratory measurements of mineral deformation and rock textures have made it possible to infer strains within the mantle part of the lithosphere that can be related to crustal deformations mapped at the surface.
- Advances in petrology and geochemistry have elucidated the pressure and temperature variations with time in rocks metamorphosed over a wide range of depths, including at subcrustal conditions. These techniques have been applied to samples brought up from depths of hundreds of kilometers in continental volcanic eruptions, allowing geochemists to constrain the history of even the deepest parts of the continental lithosphere. Based on comparisons with the results of high-pressure mineral physics experiments, some of these rocks appear to have come from as deep as the lower mantle, providing unique data on large-scale dynamical processes of the Earth's interior.

Basic research on continental processes has been applied to many of the practical problems discussed in Chapter 1, ranging from the search for natural resources to the mitigation of natural hazards. For example, field work on ancient fault ruptures (paleoseismology) has been linked with laboratory studies of rock friction and theoretical models of stress transfer to clarify how stress interactions among active faults affect the sequencing of large earthquakes (Figure 2.12). Excellent progress has also been made in understanding the shorter-term, smaller-scale processes that control earthquake nucleation and rupture dynamics. Combining this information with simulations of seismic wave propagation through complex geological structures is leading to significant improvements in the ability to forecast damaging ground motions—information crucial to seismic hazard analysis and earthquake engineering.

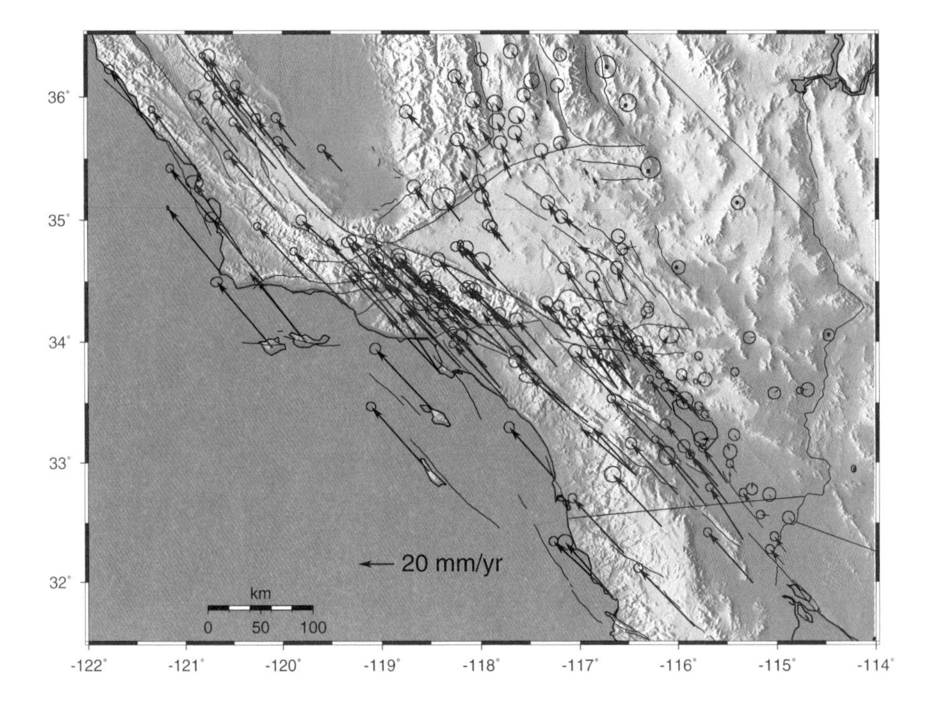

FIGURE 2.11 Geodetically determined (GPS) velocities, relative to stable interior North America, for southern California and adjacent regions of Nevada, Arizona, and Mexico. Stations located on islands in the Pacific Ocean move approximately with the Pacific plate, at ~50 mm per year northwest. In the northern portion of the diagram the zone of active strain between the Pacific plate and North America is several hundred kilometers in width, while in the southern portion of the diagram, most of the strain is localized within a zone 50 km wide and occurs on or near the San Andreas Fault. Comparison of actively accumulating deformation obtained from geodetic measurements yields important insights into seismic risk.

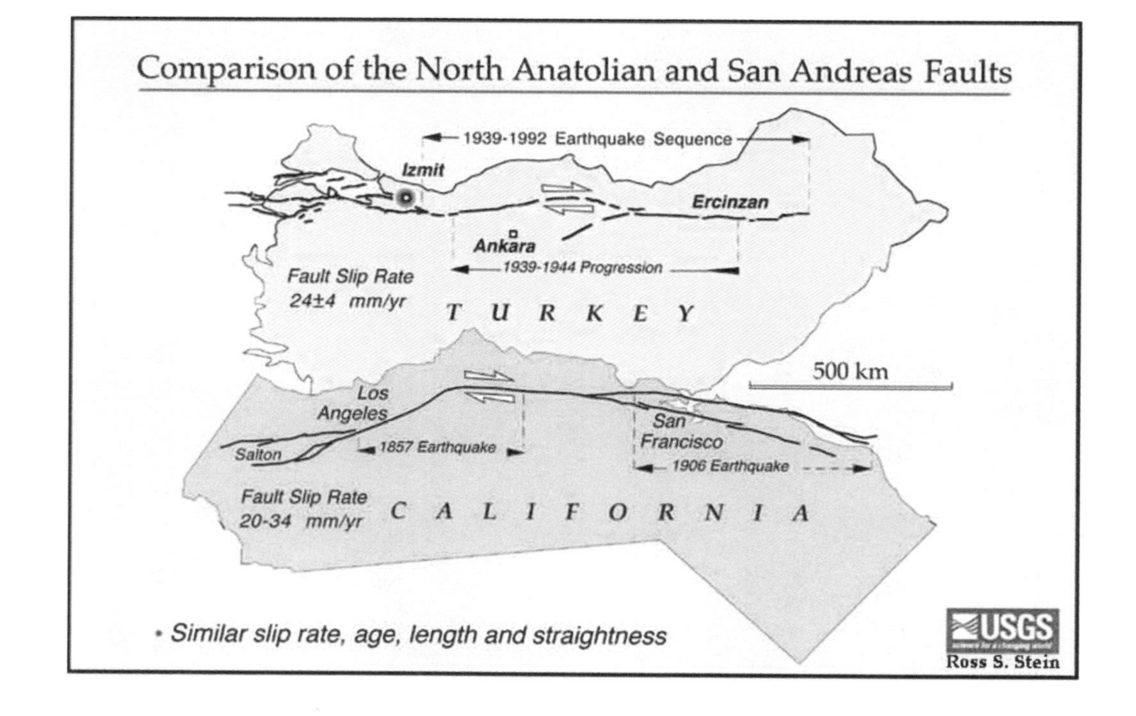

FIGURE 2.12. Comparison between major, plate-bounding active strike-slip faults in California (San Andreas Fault) and Turkey (North Anatolian Fault) showing similarities in length, slip rate, and general geometry of the fault zones. Both faults have a history of frequent earthquakes of magnitude 7 and higher, with irregular time intervals of decades to centuries between major events. The westward progression of earthquake epicenters in northern Turkey, beginning in 1939 and culminating in the Izmet earthquake of 1999, is cause for concern because the city of Istanbul (population 15 million) lies only 100 km to the west of the Izmet epicenter. Major devastation and loss of life are to be expected if earthquake activity migrates westward into this urban area. It is not known if similar patterns in earthquake migration might occur along the San Andreas Fault, but this question could be addressed through paleoseismological studies of San Andreas earthquakes over the last 10,000 years. SOURCE: R. Stein, U.S. Geological Survey.

Science Opportunities

The next decade will see discoveries derived from the rapidly expanding capabilities to observe both ancient and modern continental processes at a new level of detail. Particularly promising are the recent advances in dating capabilities and the ability to image structures deep within the Earth, allowing the temporal record obtained from geologic and geochemical studies to be correlated with ever more accurate knowledge of the Earth's three-dimensional structure.

Surface Processes, Climate, and Tectonics

What were once thought to be one-way forcing functions are now recognized to act in both directions, with feedback coupling continental tectonics, erosion processes, climate, and the composition of the atmosphere and oceans. The onset of the Indian monsoon and the desertification of sub-Saharan Africa can be related to the uplift of Tibet; the melting of midcrustal rocks and topographic collapse of mountain belts can be related to rapid denudation rates at the surface; and carbon sequestration in the oceans and atmospheres can be related to the formation and destruction of passive continental margins that are sites of carbonate deposition. These strong interdependencies are only beginning to be investigated, and the long-term evolutionary effects are very poorly known. Recent calculations suggest that current plate tectonics may be subducting up to an order of magnitude more water into the mantle than is being released out of the mantle by surface volcanism; taken at face value, the model implies that the oceans are draining back into the planet at the present time. This controversial example illustrates the unprecedented opportunities for basic research on the coupling of Earth processes from the mantle through the crust and into the oceans and atmosphere.

Active Deformation

The global forces that drive surficial motions are rooted in the mantle, but the processes of crustal deformation are governed by the ductility, fracture, and friction of rocks. These properties are not simple functions of thermodynamic state; rather, their variation is coupled in a complex way to the evolution of stress and strain. A more complete picture of the stress distribution and deformation mechanisms will require the combination of neotectonic studies with laboratory-based rock mechanics, earthquake seismology, and tectonic geodesy.[9] Particular opportunities exist for high-resolution geodetic measurements near active faults and other regions of concentrated deformation, for more detailed paleoseismic studies to gain a better history of particular fault ruptures, and for more extensive geological investigations of the way that older fault systems evolved through time. Direct measurement of fluid pressures and in situ sampling by deep drilling will be needed to understand the role of water, which is a key constituent in deformation processes.

Fluids in the Crust

Fluids play a major role in crustal processes at all depths, even in small abundances. Laboratory experiments have shown that trace amounts of water can have substantial effects on the ductile and frictional properties of rocks, underlining the importance of this ubiquitous fluid in deformation processes. The effects of fluid mobilization on chemical and thermal processes are also well established by direct observations in shallow environments, and they can be reliably inferred from high-resolution geophysical measurements and observations on metamorphic rocks uplifted from depth. The microscopic fluid-mineral interface, often modulated by biological processes (at least in the near-surface environment), governs the way in which fluids permeate and react with soils and other crustal materials. Many problems related to water and other fluids in the near-surface Critical Zone have been discussed in earlier sections of this report. Little is currently known about the dynamics of fluid conditions in the deeper parts of the crust. For example, the interior zones of well-developed strike-slip faults such as the San Andreas are thought to be much weaker than typical crustal rocks, and increased fluid pressures in the fault zone, perhaps dynamically maintained by the earthquake cycle itself, have been implicated as a possible cause of this weakness. Surface observations have been inadequate to test this hypothesis, and deep drilling will probably be required to resolve the issue. Other problems include the influence of water on the mechanics of detachment surfaces and décollements, the nature of deep circulation systems in hydrothermal areas and sedimentary basins, the fluid content of the lower crust, and the fluid flux from the mantle.

---

[9]See also *Support for Research in Tectonics at NSF*, White Paper from the Division of Structural Geology and Tectonics, Geological Society of America, July 24, 1998.

Lower Continental Crust

The lower continental crust forms a weak, ductile layer that reduces the coupling between the brittle deformation of the earthquake-producing upper crust and the aseismic motions in the mantle. Its properties, which are highly variable from place to place, play an important role in determining the width of active fault zones, postseismic response and interseismic strains, the heights of mountain belts, and the subsidence rates of sedimentary basins and passive continental margins. Surprisingly, however, some of the most basic parameters of the lower crust, such as its average composition, are still poorly constrained, and the role of key processes, such as magmatic underplating, remains largely speculative. Better knowledge of the lower crust can be obtained from geochemical and rock mechanical studies of tectonically exhumed sections, as well as from samples brought to the surface in volcanic eruptions and from geophysical research on active and ancient deformations. Substantial geological field studies in key areas will be needed to set up the tectonic framework for such investigations, and the integration of disciplinary data will require the construction of quantitative models that can account for the geomechanical behavior of the continents from the surface to the mantle.

Continental Deep Structure

Substantial knowledge has been gained about the structure of the continental lithosphere, but fundamental questions concerning its composition and formation remain unanswered. They pertain, in particular, to the anomalously thick lithosphere that underlies the Archaean cratons, which comprise the oldest parts of the crust and have acted as nuclei for continental growth. Substantial implications for Earth evolution may be drawn from the compositional and thermal contrasts among these continental keels, younger lithosphere, and the underlying convecting mantle. The keels are buoyant and not easily disrupted or subducted, and they act to organize the large-scale patterns of mantle dynamics, perhaps affecting convective processes as deep as the core-mantle boundary.

EarthScope—A Major Initiative

Building on the success of EAR's CD program, the geoscience community is working with NSF, the U.S. Geological Survey, the National Aeronautics and Space Administration, and other agencies to initiate the

EarthScope Project, a 10-year study of the North American continent (Box 2.2). A prime motivation for the project is the need for geological mapping in the third dimension, which requires imaging of the subsurface with enough horizontal resolution to observe how individual surface features—rift valleys, tectonic blocks, major faults—are expressed at depth. Essential features such as mountain belts, sedimentary basins, and province boundaries come into view only when the horizontal resolution of the images is comparable to the crustal thickness itself, about 30 km (Figure 2.13). One of the observational components of EarthScope, USArray, will take advantage of the recent advances in seismic technology that can achieve this resolution. These seismological observations will be a principal source of new structural information for modeling and interpreting a vast array of existing geological and geochemical data.

A second component, SAFOD, will extend observations in the third dimension by drilling into the San Andreas Fault, allowing direct sampling and in situ measurements of crustal rocks and fluids under conditions where the mechanics and chemistry of active deformation are very poorly known. SAFOD will provide unique data on the composition and physical properties of fault zone materials at depth, the constitutive laws governing fault behavior, the stress conditions under which earthquakes initiate and propagate, and the role of fluids in active faulting.

Two additional components, the Plate Boundary Observatory (PBO) and the InSAR initiative, will improve the geodetic observations within the North America-Pacific plate boundary zone of deformation. Continuously recording GPS station arrays and satellite-based InSAR imaging will observe small transient strains in the crust, complementing the more widely spaced GPS measurements made during the past 15 years. InSAR is a particularly promising technology that has the potential to monitor surface changes associated with a wide range of geologic phenomena and natural hazards—earthquakes, volcanoes, and landslides—as well as glacier flow and ground subsidence caused by fluid withdrawal. The InSAR component of EarthScope will increase the availability of synthetic aperture radar data to U.S. researchers, which is currently much too limited for research needs.[10]

---

[10]The amount of data that can be obtained from the European Space Agency's ERS and the Canadian RADARSAT satellites are limited, hindering application to geophysical problems. For a discussion of conditions on the use of SAR data by U.S. researchers, see *Review of NASA's Distributed Active Archive Centers*, National Academy Press, Washington, D.C., 233 pp., 1998.

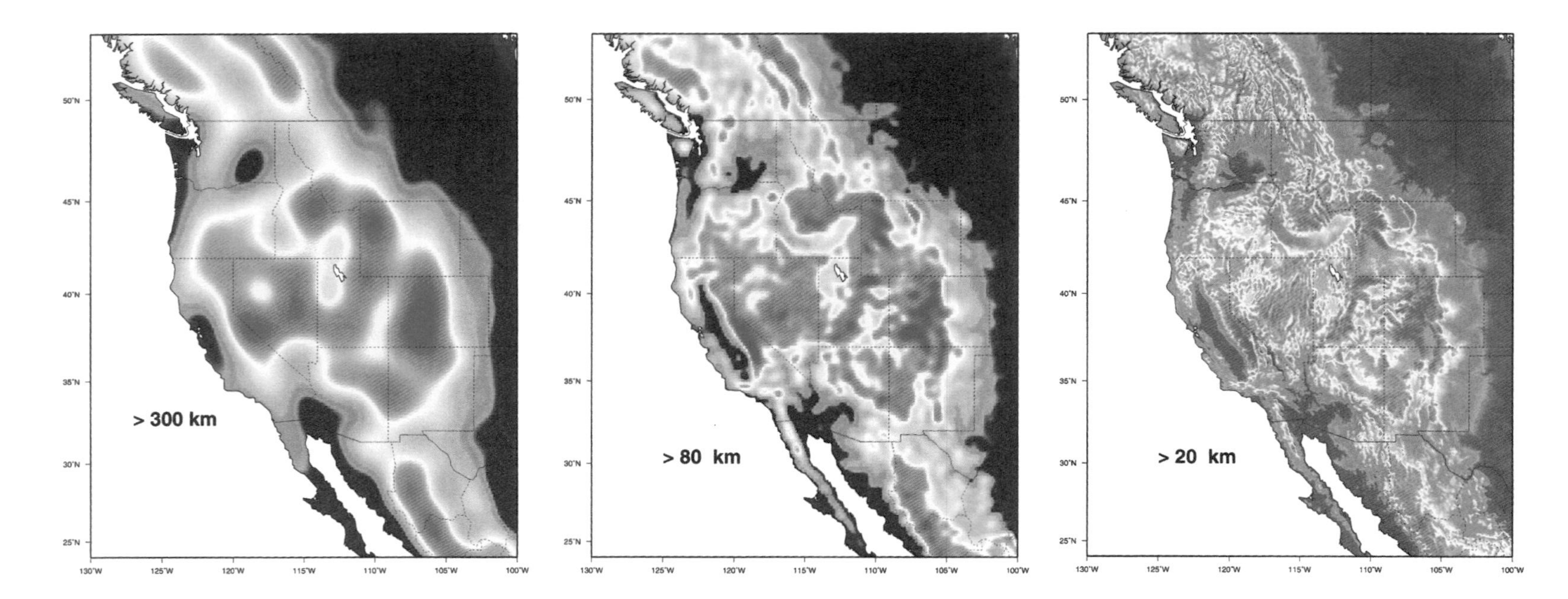

FIGURE 2.13. Topography of the broad zone of plate boundary deformation within the western United States and adjacent portions of Canada and Mexico, at low, medium, and high horizontal resolutions (300, 80, and 20 km). The height scale ranges from 400 m (violet) to 2400 m (red). At low resolution (i.e., topography smoothed to eliminate features with horizontal length scales smaller than 300 km), only the most general topographic features of this plate boundary zone can be distinguished. At medium resolution, large features such as the Sierra Nevada begin to emerge. At high resolution, small-scale mountain blocks and valleys that are key indicators of crustal extension of the Basin and Range Province appear. The low resolution in this figure is comparable to or better than current seismic tomography images of most of the upper mantle. Obtaining seismic images at the high resolution is one of the goals of the EarthScope initiative. SOURCE: G. Humphries, University of Oregon.

Box 2.2. EarthScope Initiative

EarthScope is an NSF initiative to build a network of multipurpose instruments and observatories that will significantly expand capabilities to observe the structure and active tectonics of the North American continent. The initiative will deploy four new observational facilities:

1. **USArray** will dramatically improve the resolution of seismic images of the continental lithosphere and deeper mantle below the United States and adjacent regions. USArray will have three components: (1) a transportable telemetered array of 400 broadband seismometers designed to provide real-time data from a regular grid; (2) a flexible array of ~2400 portable seismometers (using natural and explosive sources) for high-density, shorter-term observations of key targets within the footprint of the larger transportable array; and (3) a fixed network of seismometers to provide continuous long-term observations and augment the U.S. Geological Survey (USGS) National Seismic Network.
2. **San Andreas Fault Observatory at Depth (SAFOD)** will directly sample fault zone materials (rock and fluids), measure a wide variety of fault zone properties, and monitor a creeping and seismically active fault at depth. A 4-km-deep hole will be drilled through the San Andreas Fault zone close to the hypocenter of the 1966 Parkfield earthquake. Fault zone rock and fluid will be retrieved for laboratory analysis, and geophysical parameters, including seismicity, pore pressure, temperature, and strain, will be measured and monitored downhole and in adjacent areas. Instruments will be emplaced for long-term (20-year) observations of fluid activity, seismicity, and deformation.
3. **Plate Boundary Observatory (PBO)** will permit study of the three-dimensional strain field resulting from deformation along the Pacific-North American plate boundary. PBO will comprise (1) a backbone network of continuously recording, telemetered GPS receivers with 100 to 200-km spacing to provide a long-wavelength, long-period synoptic view of the entire plate boundary zone, from Alaska to Mexico, and (2) clusters of strainmeters and GPS receivers in tectonically active areas such as major faults and magmatic systems.
4. **Interferometric Synthetic Aperture Radar (InSAR)** will provide spatially continuous, intermittent strain measurements over wide geographic areas via a dedicated satellite mission to be carried out jointly between the National Aeronautics and Space Administration (NASA), NSF, and USGS. InSAR images will complement the continuous GPS point measurements of PBO. The ideal mission will provide dense spatial (100 m) and temporal (every eight days) coverage with vector solutions accurate to 1 mm over all terrain types.

SOURCE: http://www.earthscope.org.

DEEP INTERIOR

Heat from the deep interior powers convection in the Earth's liquid outer core, generating a planetary magnetic field, and in its solid mantle, driving plate tectonics and shaping the Earth's surface environment. Mantle convection is the primary generator of the Earth's topography, and it determines the geographic distribution of continents and oceans, the chemical composition of the Earth's surface layers, and the chemical fluxes into the oceans and atmosphere. The surface structure that results from these deep-seated processes is unlike that of any other planet in the solar system. Venus, comparable in size and structure to the Earth, is tectonically active, but shows no evidence of plate tectonics and is devoid of a planetary magnetic field. Recent space missions provide hints that Mars, which is much smaller than the Earth, has had plate tectonics and a planetary magnetic field in the distant geological past, although neither is currently observed. Why the Earth is so different from its planetary neighbors in terms of these global features remains the subject of considerable mystery and controversy.

Major Areas of Investigation

Mantle Convection and Geochemical Reservoirs

The flow pattern of the mantle determines the thermal history of the Earth's interior and the geological evolution of the continental and ocean floor crust. There has been no definitive means of imaging this flow pattern, however. Geophysical observations, particularly seismic tomography, provide snapshots of mantle structure, while geochemical measurements, particularly isotopic and trace element analyses, offer constraints on the past evolution of the interior; these can be related to the pattern of flow, past and present, but only indirectly. Numerical simulation of three-dimensional convection, which can now reach the parameter ranges relevant to the mantle, is proving to be an effective tool for integrating these different types of data into self-consistent models of the convection system. Laboratory studies conducted at mantle conditions provide essential constraints on the computational models and their interpretation.

Recent developments in geophysical and geochemical methods yield enormous improvements in the resolution of mantle features. In particular, seismic images now reveal that the flow driving plate tectonics, as manifested in descending lithospheric slabs, extends to the deepest regions of the mantle, at least in some areas. At the same time, geochemical data supporting the existence of two or more distinct, relatively unmixed zones of the mantle has become compelling, heightening a controversy that dates back half a century. The sharpened resolution of geophysical and geochemical observations, in combination with high-level computational simulations of the complex time-dependent flow of the mantle, is leading to new insights about the existence and long-term (billion-year) survival of distinct geological regions within the deep interior.

The Core-Mantle Boundary (CMB)

The contrasts in physical properties at the boundary between the rocky mantle and the liquid metal of the outer core are extraordinary: the mass density jumps by an amount greater than at the free surface, and the viscosity drops by more than 20 orders of magnitude—from that of "solid" rock to a value not much greater than that of water. The heat flowing across the CMB comes from energy sources within the core large enough to power a geodynamo that produces the Earth's magnetic field.

A great deal has been learned about this boundary in the last decade. Evidence of patches with highly anomalous seismic velocities ("ultralow velocity zones") and strong anisotropy (direction-dependent wave speeds), as well as indications of unexpected electrical conductivity that may influence the wobble of the Earth's rotation, are but a few of the unexpected results of recent studies. In combination with geophysical observations, laboratory investigations suggest that this is a region of intense chemical reactions, perhaps because the nature of chemical bonding is radically different at deep-Earth conditions than at the surface (e.g., oxygen, the primary constituent of rock, is a metal at the million-atmosphere pressures of the deep mantle).

The Core Dynamo and Magnetic Field

The core dynamo generates the geomagnetic field through complex electromagnetic and hydrodynamic interactions among convective motions within the rotating, highly conductive liquid outer core. Although "toy models"—simple dynamos amenable to standard theoretical analysis—have shed light on the fundamental physics of magnetic field generation, realistic simulations of the core dynamo require intense numerical calculations, and the first such numerical models were developed only about five years ago. With continuing advances in computational capabilities, these system-level models offer new possibilities for explaining observations of the geomagnetic field and core properties.

An abundance of data recently available from direct observations, satellites, and permanent observatories now reveals that the behavior of the geomagnetic field on very short time scales—from a year to a few decades—lies at the heart of understanding some of the most important processes governing its origin and temporal evolution. This is surprising, because prior work had suggested that the field evolves only on much longer periods (10^4 years), far beyond the time scale of direct monitoring. Three examples are torsional oscillations in the core, the angular momentum budget of the core, and magnetic "jerks"—changes in accelerations of the field observed on the Earth's surface.

Although the desire to understand the dynamo at a fundamental level continues to motivate studies of the geomagnetic field, there is a growing recognition that changes in the Earth's main field have important implications for a wide range of practical issues, including biological evolution, the production of carbon isotopes in the upper atmosphere by cosmic rays (essential for carbon dating), and the exchange of angular momentum between components of the Earth system. Rotation of the planet provides a dynamic link between the climate system at the surface (atmosphere and oceans) and the fluid-solid core at the center.

Inner Core

The dense, crystalline inner core, which is about two-thirds the size of the Moon, is known to play a major role in the core dynamo process. In particular, the presence of the inner core decreases the rate of change of the magnetic field and prevents the field from constantly reversing (switching north and south poles). Some numerical simulations of the core dynamo and some observations suggest that it may rotate at a rate faster than the rest of the planet. If so, observations of this "superrotation" could possibly allow monitoring of the dynamic "climate" of the overlying fluid core. The origin, growth, and subsequent evolution of the inner core remain shrouded in mystery. However, the strong heterogeneity and anisotropy in seismic wave velocities discovered in the past few years suggest that it is a tectonically active region with a rich geological history.

Disciplinary Advances

The quality and quantity of data addressing deep-Earth structure and dynamics are increasing at an extraordinary rate in several disciplines: seismology, geomagnetic studies, geochemistry, and high-pressure studies of Earth materials. As with most disciplines, important discoveries about the Earth have invariably followed the development of new instruments, deployment of new networks or application of new theoretical methods.

Seismology

The academic community undertook an ambitious program to modernize the collection of seismological data by establishing the Incorporated Research Institutions for Seismology (IRIS), initiated under NSF support in 1984. The growth in the Global Seismic Network has been phenomenal and has prompted other countries to participate in this expansion. The enlarging archive of seismographic recordings has helped to clarify the structure of the Earth's interior, substantially increasing, for example, the resolution of tomographic imaging (Figure 2.14). These images provide a snapshot view of the internal dynamics that control the geological evolution of the planet.

The USArray component of the EarthScope initiative (Box 2.2, Figure 2.15) will greatly enhance the ability to investigate Earth's interior at many scales: (1) the mantle structure beneath North America will be resolved to horizontal scales of 50-200 km down to a depth of at least 1000 km; (2) the topography and transition widths of major seismic discontinuities in the upper mantle will be resolved to tens of kilometers; and (3) the CMB structure beneath portions of the Pacific, the Caribbean, and near the Aleutians will be mapped with similar resolution. USArray will also provide the high-quality, high-density data sets needed for the joint interpretation of compressional and shear velocities.

Observatory-type studies on a planetary scale offer a number of important scientific opportunities. Large gaps exist in the global network of seismic stations that cannot be filled with island stations, particularly in the eastern Pacific and Southern Oceans. Improved spatial sampling provided by long-term, broadband seismic stations at approximately 20 ocean sites would provide much improved tomographic imaging of the structure of the lower mantle (especially in the Southern Hemisphere), the CMB, and subducting slabs and rising plumes of mantle circulation. Seafloor observatories will also provide close-in data for studies of oceanic earthquakes, which differ in substantial and poorly understood ways from continental earthquakes. A new program sponsored by the Ocean Sciences Division of NSF (Dynamics of Earth and Ocean Systems) advocates such an undertaking.

Increased computational speeds, coupled with high-speed networks and languages capable of managing parallel computations, allow calculations to be made at higher resolution and with fewer compromises than ever before. These information technologies will play a central role in the processing of very large seismic data sets and the numerical simulations of seismic wave-fields in heterogeneous, anisotropic media needed for data interpretation. These efforts could be facilitated through a better coupling between EAR and NSF information technology programs.

Isocontour Plot for SB4L18
+0.6% (blue) and -1.0% (red)
View from Top and South

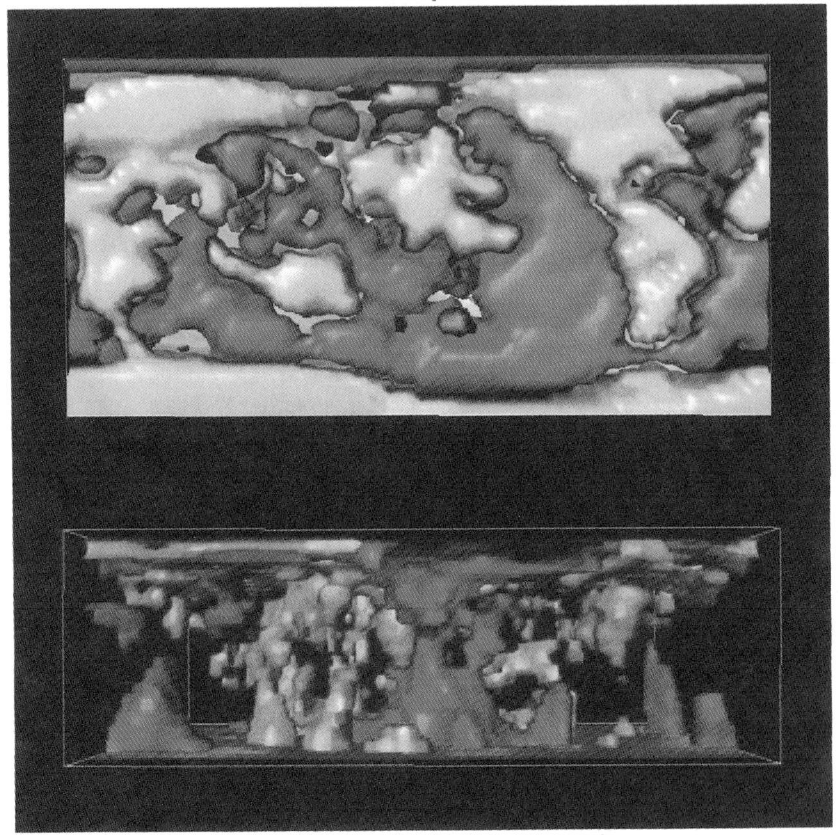

Geomagnetic Studies

High-performance computing has allowed the first realistic simulations of the convective dynamo that creates the geomagnetic field. These numerical simulations have been used to explore the possible consequences of changing thermal boundary conditions at the core-mantle boundary and have made intriguing predictions about the possible connections between the geodynamo and mantle dynamics. Analyses of the historical magnetic field show long-standing features that persist for at least hundreds of years, and comparison with the time-averaged field over 5 million years reveals intriguing similarities (Figure 2.16). The newest generation of paleomagnetic instrumentation provides measurements of an unprecedented quality. Coupled with careful field work, these new instruments permit reliable estimates of field paleointensity. Short-term variations in both direction and intensity not only furnish a means for making high-resolution stratigraphic correlations, but also place important constraints on the processes that generate the geomagnetic field. In particular, the new data can be used to develop a more detailed understanding of magnetic field reversals, and the time-averaged field can be used to understand long-term deviations from an axial dipole. Finally, extending the geographical coverage of high-resolution paleomagnetic records obtained from sediments and igneous rocks is critical for testing predictions made by core dynamo models. Drilling into high-sedimentation-rate, anoxic basins provides the best data for attaining these goals, although paleomagnetic studies of continental rocks will also yield valuable measurements.

FIGURE 2.14 Seismic tomographic image of the shear velocity of the Earth's mantle (G. Masters, G. Laske, H. Bolton, and A. Dziewonski, The relative behavior of shear velocity, bulk sound speed, and compressional velocity in the mantle: Implications for chemical and thermal structure, in S. Karato, et al., eds., *Earth's Deep Interior: Mineral Physics and Tomography from the Atomic to the Global Scale*, Geophysical Monograph 117, American Geophysical Union, Washington, D.C., p. 63-87, 2000). Shown are perturbations with respect to the spherical average, where blue and red mark isovelocity surfaces for +0.6 and -1.0%. *Top*: The map view of the shallow structure reveals seismically fast areas beneath continents and the old Pacific Ocean, whereas slow areas are found along midocean ridges. *Bottom*: View of the whole mantle from south to north (Africa to the left). Large-scale, low-velocity regions (sometimes described as megaplumes) are surrounded by fast, relatively narrow, slablike regions that extend across the whole mantle. The "ring" of fast velocities around the mid-Pacific low-velocity anomaly at the bottom of the mantle is thought to be the "graveyard of subducting slabs." SOURCE: Reprinted with permission from P. Tackley, Mantle convection and plate tectonics: Toward an integrated physical and chemical theory, *Science*, v. 288, p. 2002-2007, 2000. Copyright 2000 American Association for the Advancement of Science.

FIGURE 2.15 USArray coverage of North America including the continental margins of the United States and potential cooperating stations in Canada and Mexico. SOURCE: P. Shearer, University of California, San Diego, http://mahi.ucsd.edu/shearer/USARRAY/usarray4.html.

The current understanding of geomagnetism owes much to the geomagnetic observatories that have been maintained (in varying numbers) over the past century. Data from these observatories have been used to explore a wide range of physical phenomena, from the flow in the Earth's liquid core, to the electrical conductivity of the solid mantle, to resonant hydromagnetic oscillations in the plasma environment of the ionosphere. From these studies much has been learned, not only about the geomagnetic field per se, but also about the physical state and dynamics of the solid Earth and the near-Earth space environment. Advances in theory and computational capability, together with modern satellite data, provide exciting new possibilities for extending the knowledge (and use) of geomagnetism on a number of fronts. However, achieving this promise will require a global array of geomagnetic observatories, both on the land surface and on the seafloor, in conjunction with ongoing satellite measurements. The need for better distribution of geomagnetic observatories with modern digital equipment is a well-known problem.[11]

Geochemistry Studies

Isotope and trace element geochemistry of mantle-derived rocks, including ultramafic xenoliths and lavas erupted at midocean ridges and ocean islands, documents the chemical composition of the deep Earth, as well as the time-integrated effects of four and a half billion years of planetary differentiation. The geochemistry of such samples provides unique insights to the formation of the Earth from the solar nebula, the processes and time scales of crust and atmosphere formation, the style of mantle convection and the origin of hotspots, and the fate of slabs subducted back into the mantle. New geochemical techniques have frequently arisen from the need to analyze mantle samples for previously inaccessible elements or with enhanced precision, speed, and spatial resolution. The results of these advances have shed new light on the deep interior, and the coming decade will continue this evolution. The new generation of thermal ionization mass spectrometers, offering enhanced precision and sensitivity for radiogenic isotope analyses, has been complemented with inductively coupled plasma mass spectrometry (ICPMS), which permits routine analyses of nearly every element in the periodic table on very small samples. In the past few years, results from the first multiple-collector ICPMS instruments have also begun to appear; this new technique permits rapid isotopic analyses with minimal chemical purification as well as analyses of elements not appropriate for thermal ionization. Advances have also occurred in the ability to map small-scale (tens of microns down to submicron) chemical and isotopic variability in crystals using both laser ablation ICPMS and secondary ion mass spectrometry. Of particular interest is the ability to analyze melt inclusions, which richly document previously unknown small-scale variability in the chemical and isotopic composition of the mantle.

---

[11]For example, see *The National Geomagnetic Initiative*, National Academy Press, Washington, D.C., 261 pp., 1993.

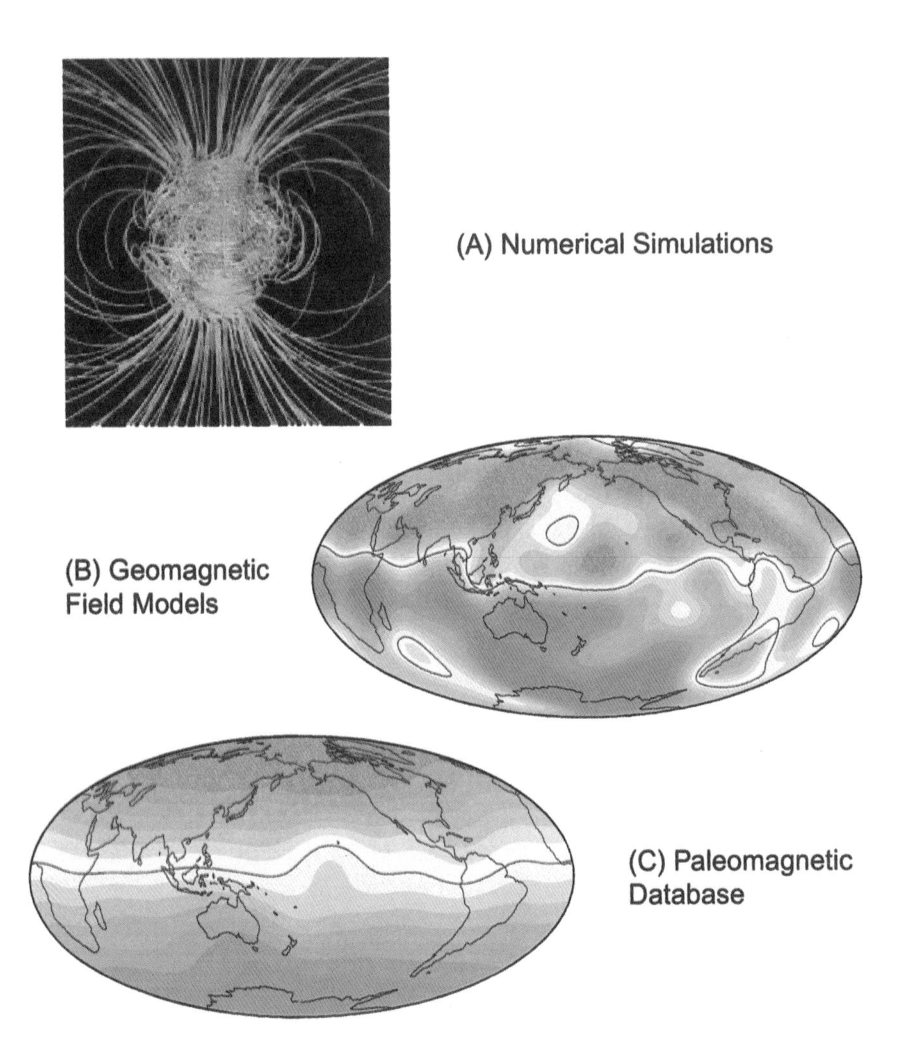
(A) Numerical Simulations
(B) Geomagnetic Field Models
(C) Paleomagnetic Database

High-Pressure Studies of Earth Materials

With the advent of dedicated synchrotron beamlines and other modern technologies, it has become possible to document the properties of Earth materials in situ, at the high pressures and temperatures of the planet's deep interior. Measuring the elastic properties that determine seismic wave velocities; the density, thermal conductivity, and rheological properties that control geodynamic motions; and the partitioning of minor and trace elements that produce the geochemical signatures observed in rock samples is now possible at deep-Earth conditions. First-principles quantum and statistical mechanical calculations are also offering significant theoretical insights into the chemical and physical properties of materials throughout the mantle and core. Thus, the seismic anisotropy of the inner core, the possible chemical and physical interactions at the core-mantle boundary, the distribution of radioactive heat-producing elements, and the initiation of melting (or crystallization) throughout the Earth are all subject to quantitative study over the coming decade. The abundance and distribution of water and other "volatile" molecules within the planet, as well as their cycling between the surface and interior over geological history, can be evaluated for the first time. These are specific examples of the broader question now being addressed through high-pressure studies; What is the current state of the Earth's interior, and what are the processes by which it evolved to this state?

FIGURE 2.16 (A) Results of numerical simulations of the geomagnetic field of Glatzmaier and Roberts. Blue lines indicate flux into the core, and orange is outward-directed flux. External to the core, the field is approximately that of a dipole. Numerical simulations make predictions about field behavior (average intensity, secular variation, behavior during reversals) for different assumed boundary conditions. SOURCE: G.A. Glatzmaier and P. Roberts, A three-dimensional self-consistent computer simulation of a geomagnetic field reversal, *Nature*, v. 377, p. 203-209, 1995, Reprinted by permission from *Nature*. Copyright 1995 Macmillan Magazines Ltd. (B) Field model UFM1 of Bloxham and Jackson. Averages of geomagnetic field observations over the last 300 years are plotted as radial flux on the core mantle boundary. Color conventions are as in (A). Although the field is dominantly dipolar, there are significant departures (flux patches with the wrong color) from a dipole model that persist for hundreds of years. SOURCE: J. Bloxham and A. Jackson, Time-dependent mapping of the magnetic field at the core-mantle boundary, *Journal of Geophysical Research B*, v. 97, p. 19, 537-19,563, 1992. Copyright 1992 American Geophysical Union. (C) Averages of paleomagnetic data spanning the last 5 million years. There is loss of resolution relative to the geomagnetic field average shown in (B) due to the difficulty in getting global coverage with high-quality data. SOURCE: Reprinted by permission from C.L. Johnson and C. Constable, The time-averaged geomagnetic field; Global and regional biases for 0-5 Ma, *Geophysical Journal International*, v. 131, p. 643-666, 1997. Copyright 1997 Geophysical Journal International.

THE PLANETS

The Earth is only one member of a rapidly growing family of known planets, both within and beyond the solar system. The extraordinary pace of planetary exploration, driven by large national and foreign investments in spacecraft and telescopes, is expected to continue for at least the next decade. Current or planned space missions will provide unprecedented detail and coverage of the geology, topography, structure, and composition of many solar system bodies; in several cases this coverage will rival or exceed equivalent terrestrial data. Similarly, it is anticipated that by 2010 the first samples collected from Mars, a comet, an asteroid, and the Sun (via solar wind particles) will be returned to Earth for direct investigation of their composition and structure. Telescopic observations of primitive objects in the solar system (i.e., Kuiper belt objects, located beyond the orbit of Neptune) and of the planets orbiting distant stars promise to provide unique data regarding the origin and evolution of our solar system.

Investigations of the solid Earth have usually been undertaken through studies of the Earth itself. However, the study of extraterrestrial materials has provided some of our most important insights regarding the Earth. For example, the age of the Earth, 4.56 billion years, was derived not from dated Earth material but by the study of meteorites (and subsequently confirmed by dating lunar samples). The baseline for discussions of the bulk composition of the Earth is likewise based on a meteorite (chondrite) reference. The richness of materials and information on extraterrestrial bodies ensuing from the next decade of exploration will provide important new opportunities for basic research into the origin, evolution, and structure of planets, including Earth. Effective use of these new data will require both a broad-based effort to promote interaction between the Earth science and planetary science communities, and a substantial enhancement of analytical capabilities.

Promise of Planetary Exploration

Robotic exploration of the solar system is increasing rapidly. The scientific opportunities provided by these missions are detailed in a number of National Research Council reports.[12,13] Here, the committee focuses on near-

---

[12]*An Integrated Strategy for the Planetary Sciences: 1995-2010*. National Academy Press, Washington, D.C., 199 pp., 1994.

[13]*The Exploration of Near-Earth Objects*. National Academy Press, Washington, D.C., 32 pp., 1998.

term missions of interest to the solid-Earth sciences. Within the decade, current or planned NASA or international missions will provide the following:

- *Surface topography, gravity, and magnetic field data of unprecedented resolution for the terrestrial planets and for some satellites*: For example, the Messenger Orbiter will define Mercury's gravity field and Cassini flybys will refine the gravitational parameters of some Saturnian satellites. High-resolution geophysical observations, such as those collected by the Mars Global Surveyor (MGS) mission, are critical for identifying and understanding the geologic processes that shape solid planets. While much is known regarding these processes on Earth, existing data leave little doubt that conditions on other planets and satellites can yield fundamentally different results (Figure 2.17).

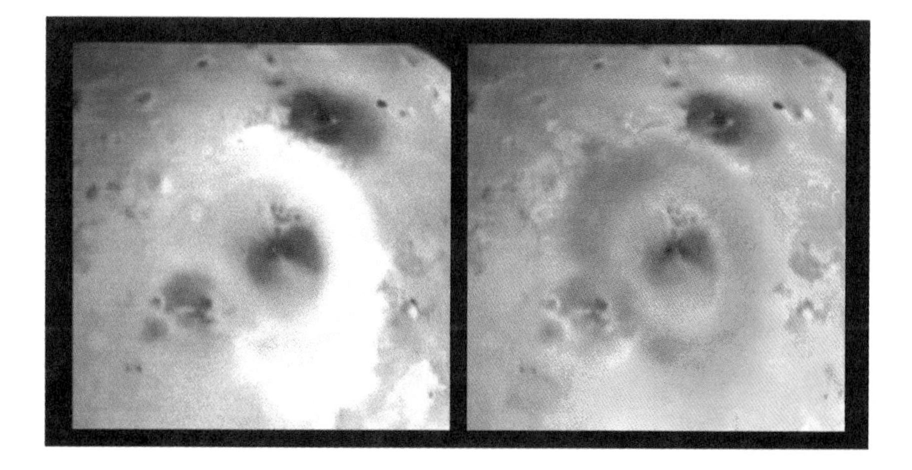

FIGURE 2.17 Active volcanism on Jupiter's moon Io from the Galileo spacecraft. The true-color image on the left shows the ~1300-km-diameter red ring of sulfur surrounding the volcano Pele. The false-color infrared composite on the right reveals the glow of hot (up to 1027°C) magma in the volcano's central crater. SOURCE: NASA Jet Propulsion Laboratory.

- *In situ chemical and mineralogical analyses of the surfaces of Mars, satellites, asteroids, comets, and the solar wind* (complemented by samples returned from these same sources): These data will open an exciting new research frontier documenting and understanding chemical processes that produced the solar system and that modify both the surfaces and the interiors of solar-system bodies. Such work was largely impossible from the earlier generation of orbiting spacecraft. The results will be important in a wide array of Earth and planetary science disciplines. Some key research areas benefiting from these observations include the composition and early evolution of the solar system, the climatic history of Mars, and the possible existence of life beyond the Earth.

In addition, data collected from ongoing missions (e.g., the surprising diversity of Jovian satellite magnetic fields detected by the Galileo mission) will be analyzed over the next several years. A lengthy period of analysis of satellite data, combined with other geophysical observations, is important for understanding the chemical and physical structure and dynamics of planets.

Science Opportunities

Continuing exploration of the solar system, particularly Mars, will provide both new opportunities and challenges for the Earth science and planetary science communities in the coming decade (Box 2.3). On the one hand, new high-resolution observations of key physical, geological, chemical, and mineralogical characteristics will permit insights to the processes occurring on and in other solar system bodies. As the quantity and quality of observations approach those for the Earth, the knowledge-base within the Earth science community will become increasingly important for interpretation and comparison. On the other hand, these observations will also contribute to an improvement in our understanding of Earth and the solar system as a whole:

- Geological and geophysical processes occurring on other planets are responses to the same basic forces as on Earth, but applied in different ways to materials with different properties and subject to different boundary conditions. As such, these systems will provide new environments for investigating basic planetary phenomena.
- Unlike the Earth, which is continually resurfaced and eroded, many bodies preserve a physical and chemical record from the earliest days of the solar system. Thus, they provide data regarding planetary evolution that no longer exists on Earth. For many fundamental processes that occur early in a planet's history (e.g., formation of the first crust), the Earth's record is extremely difficult to decipher.
- Distinctive chemical and isotopic composition is a first-order property of the solar system, and the new data likely to be obtained over the next

decade are critical for furthering understanding of the processes of mixing, accretion, and differentiation of planets and meteorite parent bodies.

- Processing local resources to obtain various materials will be crucial to the establishment of self-sufficient planetary bases.[14] The study of soil formation and past aqueous weathering on planetary surfaces will be a prerequisite for developing extraterrestrial "soils" for growing plants. The factors that influence the activity, ecology, and population dynamics of microbes in soils are important for nutrient cycling, biodegradation, and regenerative life support on a planetary base.

Box 2.3. Recent Studies of Mars

NASA has identified exploration of Mars as a top research priority, and early results of this focused effort are now being obtained. These results illustrate some of the basic research opportunities and challenges posed by intensive solar system exploration:[1,2]

- Much of the interest in Mars has arisen from the study of SNC (Martian) meteorites, which permit detailed studies of rocks derived from a planetary body similar to Earth. In addition, one of these meteorites was initially believed to carry evidence of fossil life. Although the existence of fossils is not widely accepted, the exciting possibility of detecting and characterizing extraterrestrial life (or its fossil imprints) has invigorated study of the physical, chemical, and mineralogical composition of Martian rocks in a way analogous to what future returned samples will require. The lesson from these studies is that understanding detailed aspects of another planet (e.g., composition of the planet, petrogenesis and planetary evolution, climate history, possible existence of life) will require intensive and diverse studies with investigators derived from the planetary science and Earth science communities, and beyond. For studies at the micro- and nanoscale and with very limited amounts of material, new analytical techniques and instrumentation are critical. This broad effort to understand Mars has required specific initiatives from federal funding sources (i.e., NASA, NSF) because existing mechanisms within these organizations were not suited to such an effort.
- In 1998 the Mars Global Surveyor began orbiting Mars, and the first science returns were reported in early 1999. As with the SNC meteorites, the data already obtained from this mission illustrate important new linkages between the planetary and Earth sciences. For example, the first high-resolution magnetic survey of Mars has revealed regions of alternating magnetic polarity that provide exciting and unexpected insight to

---

[14] *Microgravity Research in Support of Technologies for the Human Exploration and Development of Space and Planetary Bodies*, National Academy Press, Washington, D.C., in press, 2000.

the evolution of the Martian crust and interior (Figure 2.18). That is, the strength and variability of magnetization demonstrate the unexpected existence of an Earth-like planetary dynamo within the core of Mars. Because Mars no longer has a strong magnetic field, its planetary dynamo must have ceased to operate sometime after the formation of the crust. These observations raise important questions about why the Martian dynamo behaved so differently from that of the Earth. In addition, the magnetizations have been interpreted as the equivalent of terrestrial "seafloor stripes" that are the hallmark of plate tectonics. Although this interpretation is controversial, the data are of first-order importance. Further effort in understanding how planetary dynamos and plate tectonics might work on Mars in comparison with the Earth is critical for accurately interpreting and understanding these data.

- Another important area being explored by MGS is the climate history of Mars. Photographic evidence has long suggested features derived from flowing water on Mars, but recent high-resolution images show unexpected complexity, raising questions regarding how the features were produced. Whereas some features look like flood deposits and channels, others look like subsurface flow. Further research using photogeologic techniques and other methods is necessary for understanding how flowing water under conditions very different from those on the Earth affects the surface morphology of planets. Similarly, recent data on surface mineralogy furnishes limited evidence for pervasive aqueous alteration of Martian soils. This clearly has bearing on the longevity of Martian surface waters, a key question for the history of climate and possible life on Mars. Additional work is necessary to understand basic soil formation and alteration processes.

[1] *An Integrated Strategy for the Planetary Sciences: 1995-2010*. National Academy Press, Washington, D.C., 199 pp., 1994.

[2] *Review of NASA's Planned Mars Program*. National Academy Press, Washington, D.C., 29 pp., 1996.

Analytical Challenges

In situ planetary studies will offer great benefits for Earth and planetary scientists,[15] but sample return offers the most promise for research funded by EAR. Samples brought back from various solar system bodies will share a common trait—the amount of material will be extremely small, ranging from perhaps a kilogram of Martian soil and rock to submilligram quantities of comet dust. In comparison, hundreds of kilograms of lunar materials were returned during the Apollo program. The small sample sizes, coupled with the

[15] *A Scientific Rationale for Mobility in Planetary Environments*. National Academy Press, Washington, D.C., 56 pp., 1999.

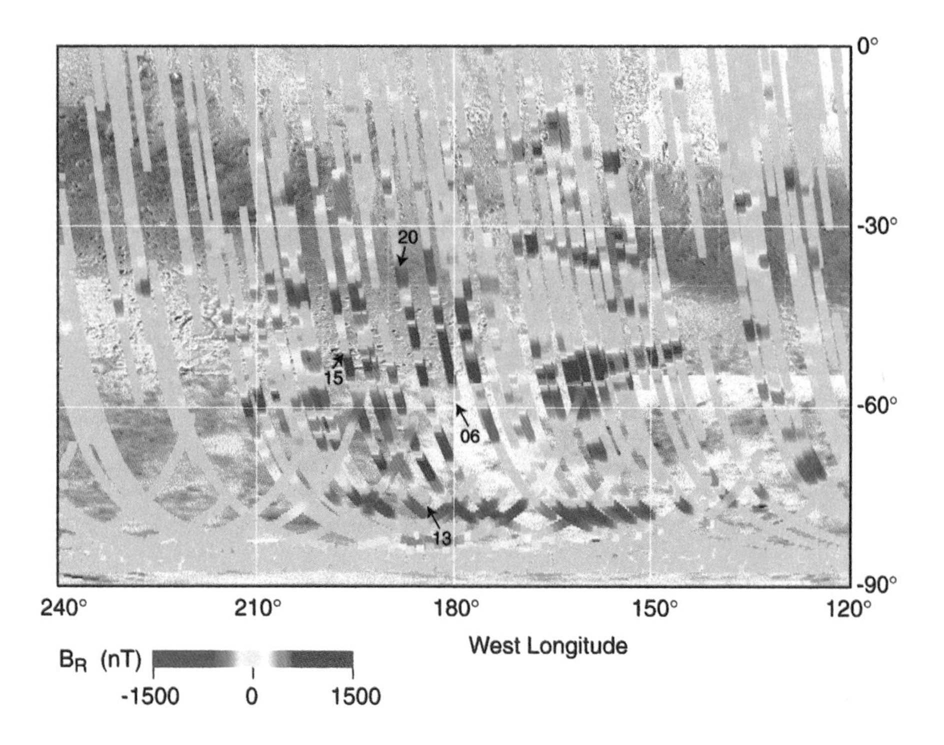

FIGURE 2.18 Clues to the ancient magnetic field of Mars from the Mars Global Surveyor spacecraft. This image documents the strength and variability of magnetization recorded by Martian crustal rocks. The pattern indicates the existence of an Earth-like planetary dynamo operating in the Martian core, but only very early in the history of the planet. The alternating polarity pattern records fundamental, yet previously unknown aspects of the formation of the early Martian crust, possibly by a process similar to seafloor spreading on Earth. SOURCE: Reprinted with permission from J.E.P. Connerney, M.H. Acuna, P.J. Wasilewski, N.F. Ness, H. Reme, C. Mazelle, D. Vignes, R.P. Lin, D.L. Mitchell, and P.A. Cloutier, Magnetic lineations in the ancient crust of Mars, *Science*, v. 284, p. 794-798, 1999. Copyright 1999 American Association for the Advancement of Science.

need to investigate the samples for a tremendous diversity of purposes, place a new and extremely strict requirement for high-efficiency analyses. In addition, many types of analyses will require extremely high spatial resolution. For a number of critical measurements, present techniques and instruments are simply not adequate.

An important challenge will be to construct appropriate instrumentation for these analyses. The necessary instrumentation is diverse, and includes advanced ion microprobes, mass spectrometers, and electron microscopes, as well as accelerator- and synchrotron-based analytical probes. This challenge, if met, is an important opportunity for the Earth science community. As evidenced by the blossoming of isotope geochemistry following the instrumentation campaign associated with the Apollo program, the creation of new techniques and capabilities can invigorate an entire field for many years.

3

Findings and Recommendations

Basic and applied research in Earth science is guided by an ambitious scientific program to understand the Earth and its history through the multidisciplinary study of the dynamics and evolution of terrestrial systems. Chapter 1 discussed how societal needs drive research on system-level problems and how this approach is being enabled by across-the-board improvements in observational capabilities and information technologies. Chapter 2 reviewed the status and prospects of basic research in six important problem areas spanning a wide range of future activity in Earth science:

1. integrative studies of the Critical Zone, the heterogeneous, near-surface environment, where complex interactions involving rock, soil, water, air, and living organisms regulate natural habitats and determine the availability of life-sustaining resources,
2. geobiology, which addresses the interactions of biological and geological processes, the evolution of life on Earth, and the factors that have shaped the biosphere,
3. research on Earth and planetary materials, which uses advanced instrumentation and theory to determine properties at the molecular level as the basis for understanding materials and processes at all scales relevant to planets,
4. investigations of the three-dimensional structure and composition of the continents, the geologic record of continental formation and assembly, and the physical processes in active continental deformation zones,
5. studies of the Earth's deep interior to define better its structure, composition and state and to understand the machinery of mantle convection and the core dynamo, and

6. planetary science using extraterrestrial materials, as well as astronomical, space-based, and laboratory observations, to investigate the origin, evolution, and present structure of planetary bodies, including the Earth.

This chapter presents the findings and recommendations that have been drawn by the committee from its overview of the science opportunities and societal needs. In constructing its recommendations, the committee was cognizant that the National Science Foundation (NSF) must continually strive to balance its funding of basic research among (1) core programs that support investigator-driven, disciplinary activities; (2) problem-focused programs of multidisciplinary research; and (3) equipment-oriented programs for developing new instrumentation and facilities. It is the committee's conclusion that the Earth Science Division (EAR) has done an excellent job at maintaining such a balance in the past. The committee therefore offers recommendations relevant to all three programmatic areas that, if implemented, will address the science requirements for the next decade. It also comments on opportunities for coordinating EAR-sponsored research with programmatic activities in other NSF divisions and with other agencies.

LONG-TERM SUPPORT OF INVESTIGATOR-DRIVEN SCIENCE

It is commonly believed that many of the most significant conceptual breakthroughs in science come at the hands of individual investigators or small groups of researchers, rather than through structured collaborations. As indicated by the workshop reports, letters to the committee, and the 1998 report of the EAR visiting committee, the Earth science community supports the notion that individual investigators should be free to pursue their own research directions. One letter from an Earth scientist put it succinctly: "Too much emphasis on 'large' science at the expense of 'small' science will, over time, stifle creativity." It is particularly important that creative young scientists be allowed to conduct scientific research of their own conception, rather than projects conforming to the current scientific consensus, which is often articulated by established groups (such as National Research Council [NRC] committees). The committee strongly endorses this point of view.

Finding: EAR funding of research projects initiated and conducted by individual investigators and small groups of investigators is the single most important mechanism for maintaining and enhancing disciplinary strength in Earth science.

EAR is currently divided into five disciplinary core programs based on divisions among the constituent fields of Earth science—Geology and Paleontology (G&P), Geophysics, Hydrologic Sciences, Petrology and Geochemistry, and Tectonics—and three core programs that are intrinsically multidisciplinary—Continental Dynamics, Education and Human Resources, and Instrumentation and Facilities (see Appendix A). Proposals from individuals or small groups of investigators are funded primarily through the first five. The strength of these core programs is thus critical to sustaining advancement in all aspects of Earth science, in particular new avenues of research and multidisciplinary investigations of Earth processes and systems.[1] Core funding of individual investigators also underpins research-based education at NSF (see "Education" below), as well as the continuing education of recent graduates who must learn the value of innovation and independence to be successful researchers in their own right.

Strict disciplinary divisions are recognized to be artificial, and an increasing number of investigator-initiated "small-science" projects span two or more disciplines (see Chapter 2). For this reason, NSF program managers can and do work together to evaluate and split-fund proposals. Flat budgets and declining buying power within the disciplinary core programs, as well as the sometimes narrow focus of review panels, have made it increasingly difficult to accommodate new areas of investigation within this structure, however.[2] The problem is particularly acute in two fields identified as exceptionally promising in Chapter 2—geobiology, and Earth and planetary materials—for which major new support is justified. Outstanding research opportunities related to the multidisciplinary problems of the Critical Zone also warrant expansions in the traditional fields of geology and hydrology.

---

[1]In their recent essay on the problems of interdisciplinary research, Norman Metzger and Richard Zare (*Science*, v. 283, p. 642-643, 1999) put forward this point in the following way: "Strong interdisciplinary programs can only be built in circumstances in which strong disciplinary programs already exist. It makes no sense to sacrifice successful disciplinary efforts to appease perceived interdisciplinary needs."

[2]According to the budget figures presented in Appendix A, the total 1999 expenditure in the EAR disciplinary core programs was $38 million, compared to $39 million in 1992 and $33 million in 1985. In terms of constant buying power, this represents a 23% decline in the funding of disciplinary core programs between 1985 and 1999. On the other hand, funding of other EAR program elements, including multidisciplinary research, instruments, facilities, and science and technology centers, more than tripled over the same period (see Figure A.1).

Geobiology

Geobiology encompasses research on the interactions among biological, geological, and environmental processes on the evolution of life on Earth, and on the factors that have shaped the long-term evolution of both life forms and their environment. The development of new and powerful tools in the biological sciences (genomics, proteinomics, and developmental biology) and in geochemistry, mineralogy, stratigraphy, and paleontology offers unprecedented opportunities for advances. Two primary research directions are identified:

1. *Topics related to interactions between Earth systems and biologic processes*:
   - How the habitability of the Earth is affected by natural and anthropogenic environmental change
   - Function and diversity of microbial life in a wide range of environmental systems, and the relative importance of biological and inorganic processes in weathering, soil formation, mineralization, and other geological and pedological processes
   - Biogeochemical interactions and cycling among organisms, ecosystems, and the environment, including applications of geomicrobiology to monitoring and remediating environmental degradation
   - Relationship between ecology and climate change, including the role of rare events in reshaping ecosystems and climate, from local to global scales

2. *Topics related to the origin and evolution of life on Earth*:
   - Biological and environmental controls on how species diversity changes through time, including ecological and biogeographic selectivity, and causes of extinction and survival
   - Nature of evolutionary innovation through the integration of biological, fossil, and geochemical data
   - Rates at which organisms, communities, and ecosystems are able to respond to environmental perturbation over short and long time scales.

None of the existing core programs have the intellectual scope or sufficient resources to accommodate a prolonged emphasis on geobiology. The most closely related program within EAR is G&P, but it is already severely oversubscribed[3] and is thus unable to adequately cover many important biological and geochemical aspects of geobiology.

---

[3]Over the past five years, G&P received approximately 290 proposals per year; its success rate was 20-25%, compared with an overall EAR success rate of 31%.

Recommendation: EAR should seek new funds for the long-term support of geobiology to permit studies of the interactions between biological and geological processes, the evolution of life on Earth, and the geologic factors that have shaped the biosphere.

A geobiology program would be the basis for a powerful scientific partnership between Earth sciences and biological sciences, directed toward a comprehensive understanding of the relationship between life and its planetary environment. A successful program would integrate geological, paleontological, environmental, geochemical, pedological, oceanic, atmospheric, and many types of biological data. It would also add key biological perspectives to research initiatives in Earth and planetary materials, discussed below. Thus, it will be crucial to forge strong links between geobiology and existing programs within NSF, including the Ocean Sciences Division (OCE) (biogeochemical interactions between marine organisms and their environment); the Environmental Geochemistry and Biogeochemistry (EGB) Program (biogeochemical processes in the near-surface environment); the Earth System History (ESH) Program (ecology and climate change); Life in Extreme Environments (LExEn) (microorganisms in extreme environments and the origin of life), Water and Energy: Atmospheric, Vegetative, and Earth Interactions Program (Earth's hydrologic and energy cycles); and the Biological Sciences Directorate (evolution of ecosystems, function of organisms as a function of their environment, microbial biology and growth in natural environments, and microbial processes in energy flow and nutrient cycling). Partnerships with other agencies will also be important, particularly the Department of Energy (DOE) (organic geochemistry, carbon cycle and bioremediation); U.S. Geological Survey (USGS) (surficial cycles and processes, and effects on contaminants on organisms); the Environmental Protection Agency (EPA) (environmental biology and microbes in the environment); the U.S. Department of Agriculture (USDA) (greenhouse gas budgets, microbial cycling of elements, soil development); the National Institutes of Health (NIH) (molecular biology and soil-borne pathogens); and the National Aeronautics and Space Administration (NASA) (astrobiology, response of terrestrial life to environmental change). A geobiology program would also sponsor a needed Earth science component to the new NSF initiative in Biocomplexity[4]—which is in many ways concerned about the central issues of "geocomplexity"—as well as those that may derive

---

[4]The Biocomplexity Initiative is an NSF-wide multiyear effort to understand the nature and dynamics of biocomplexity in the environment. The first phase will focus on the functional interrelationships between microorganisms and biological, chemical, geological, physical, and social systems. See http://www.nsf.gov/home/crssprgm/be/.

from the recommendations of the National Science Board's Task Force on the Environment. Substantial effort will be needed to undertake the cross-disciplinary training required by geobiological inquiry (see "Education" below).

Earth and Planetary Materials

Research on Earth and planetary materials has emerged as a field distinct from, but highly complementary to, the well-established disciplines of geochemistry, geophysics, and petrology that are currently among the core programs at EAR. The major research domains encompass mineral physics, planetary materials research and their interfaces with geomicrobiology, soil science, and biomineralization (including nanocrystalline phases), rock physics, mineral and rock magnetism, and the science of mineral surfaces. The Earth and planetary materials community is identified through its use of distinct tools and methodologies characterized by an atomistic approach and its reliance on major experimental facilities. It serves the geophysical, geochemical, and geological communities, yet shares a close kinship with communities in chemistry, materials science, and condensed-matter physics. The laboratory-based measurements by researchers in this community are essential for the quantitative interpretation of observations of the Earth and planetary bodies made by geochemists, geophysicists, mineralogists, petrologists, and soil scientists.

The committee has identified a number of opportunities for new research emphases in Earth and planetary materials:

- biomineralization (natural growth of minerals within organisms, as well as applications to the development of synthetic analogues),
- characterization of extraterrestrial samples (e.g., returns from Mars, comets, and interplanetary space),
- superhigh-pressure (terapascal) research, with applications to planetary and stellar interiors,
- nonlinear interactions and interfacial phenomena in rocks (e.g., strain localization; nonlinear wave propagation; fluid-mineral reactions, whether with aqueous fluids, magmatic fluids, or both; and coupling of chemical reactions and fracturing),
- nanophases and interfaces, including microbiology at interfaces (e.g., catalyzing or modulating geochemical reactions) and applications to the physics and chemistry of soils,
- quantum and molecular theory as applied to minerals and their interfaces (e.g., mineral interactions with fluids and gases) and aggregates (rocks and soils), and

• studies of granular media, including the nonlinear physics of soils and loose aggregates.

The potential for initiatives in these areas has been increased by numerous new collaborations of geoscientists with chemists, biologists, physicists and materials scientists.[5] Past successes in applying major facilities, such as synchrotron beamlines, argues for the community's leading in the development of new facilities (e.g., new-generation neutron and X-ray beamlines and micro-analytical tools) and their application to Earth, environmental and planetary problems. These are just becoming feasible, and their development, enhanced by recent major advances in theory, is being pursued in associated disciplines.

Recommendation: EAR should seek new funds for the long-term support of investigator-initiated research on Earth and planetary materials to take advantage of major new facilities, advanced instrumentation and theory in an atomistic approach to properties and processes.

A research program in Earth and planetary materials would offer a new mechanism to create more substantial linkages with existing programs outside EAR, including in NSF's Astronomy Division (studies of planetary origins and interiors; superhigh-pressure research) and Division of Materials Research (research on complex oxides and other mineral-like materials), DOE (e.g., Basic Energy Sciences, Biological and Environmental Research, and Environmental Management), and NASA (e.g., Planetary Geology and Geophysics, and Cosmochemistry). Moreover, the scientists that study Earth and planetary materials have been at the forefront of research on nanomaterials and are well positioned, therefore, to contribute to the interagency National Nanotechnology Initiative.[6]

Hydrology, Geology, and the Critical Zone

Integrative studies of the Critical Zone will depend on strong disciplinary programs in soil science, hydrology, geobiology, sedimentology, stratigraphy,

---

[5]*Microscopic to Macroscopic: Opportunities in Mineral and Rock Physics and Chemistry*, results of a workshop held in Scottsdale, Arizona, May 28-30, 1999.

[6]The multiagency National Nanotechnology Initiative is concerned with the creation of useful materials, devices, and systems through the control of matter on the nanometer scale and the exploitation of novel properties and phenomena at that scale. See http://www2.whitehouse.gov/WH/EOP/OSTP/NSTC/html/committee/ct.html.

geomorphology, and coastal zone processes. Currently, research in these diverse fields is funded mainly through the G&P and Hydrologic Sciences core programs.

Hydrology

In hydrology, many of the advances of the last two decades have resulted from the increasing ability to measure with greater spatial and temporal resolution the fluxes and states that are critical for quantifying the water balance. Enhancements in computer and information technology have made possible, for example, the development and testing of distributed catchment models using remote-sensing information at scales ranging from tens of meters to tens of kilometers. Better characterization of the heterogeneous nature of subsurface flow has led to an improved understanding of the movement of groundwater and the fate of contaminants. Knowledge of the nature and severity of floods and droughts has increased, although much remains to be done particularly in connection with discerning human influences on the global energy and water balances and in forecasting hydrologic extremes.

Understanding and quantifying the various chemical cycles of the Critical Zone require extremely accurate hydrologic balances for both very short and very long time spans. New efforts are required to measure and model the water pathways through vegetation and the vadose zone into the groundwater and below, as well as to determine the corresponding biogeochemical interactions and define human influences on hydrologic and ecologic systems. These efforts are needed in a variety of geologic, soil, climate, and vegetation settings and for spatial scales that range from a few meters at the hillslope scale to basin-wide and continental scales.

Recommendation: Owing to the significant opportunities for progress in the understanding of hydrologic systems, particularly through coordinated studies of the Critical Zone, EAR should continue to build programs in the hydrologic sciences.

Increased support of the hydrological science program would improve the success rate of proposals (19%) and complement the focused, largely intramural hydrology programs in the National Oceanic and Atmospheric Administration (NOAA) (e.g., the Global Energy and Water Cycle Experiment [GEWEX] Continental-Scale International Project) and the USGS Water Resources Division, as well as the remote-sensing hydrology programs in NASA (GEWEX, Land Surface Hydrology Program).

In some cases, the scientific problems will be best addressed by establishing long-term natural laboratories in specific climatic and land-use settings. The opportunities for Earth science research in natural laboratories are discussed further under "Major Initiatives." Contemporaneous investigations at diverse locations will be needed to ascertain the large-scale, climate-driven hydrologic connections that bear directly on the Critical Zone. The USGS, USDA, NOAA, and NASA will be important partners for EAR. The proposed Second International Hydrologic Decade may furnish significant opportunities for exercising this partnership. The primary goals of the decade—to reverse the decline in the development of hydrologic observing systems and to reduce uncertainties in the measurements—bear directly on Critical Zone research on flood and drought hazards and coupled ecologic-hydrologic systems.

Geology

The G&P program sponsors a wide range of research on physical, chemical, and biological processes that take place in the Critical Zone (Appendix A). This emphasis on the Critical Zone would be further strengthened by moving paleontology from G&P to Geobiology. Such a reorganization would also free up funds to address scientific problems that are related to the Critical Zone, but are not easily funded within the current program structure. For example, studies related to soils and coastal zone processes have received relatively little attention to date, in part because the scope of the research spans the Geosciences Directorate. For example, sedimentary and geochemical fluxes in the near-shore environment are important indicators of Critical Zone processes, but NSF programmatic areas typically stop at the shoreline—the marine aspects are funded through OCE and are thus separated from the sedimentary aspects, which must vie for support from EAR.[7] Similarly, soils constitute a major reservoir of carbon, but studies on the release and sequestration of carbon and other greenhouse gases are funded primarily through the Atmospheric Sciences Division (ATM) and OCE, divisions that have little expertise in soils. EAR, on the other hand, is only beginning to recognize the contributions that soil scientists can make to the study of the Earth.

Recommendation: EAR should enhance multidisciplinary studies of the Critical Zone, placing special attention on strengthening soil science and the study of coastal zone processes.

---

[7]*Coastal Sedimentary Geology Research: A Critical National and Global Priority*, results of a workshop held in Honolulu, Hawaii, November 9-12, 1999.

The committee recognizes that much work is needed to implement a research program on the Critical Zone that is sufficiently visionary and broadly based to address the wide range of fundamental and applied problems involved in the study of the near-surface environment. In the long term, EAR will have to coordinate its programs with ATM (paleoclimate and trace gas fluxes), OCE (geochemical pathways at the ocean-solid-Earth interface, paleoceanography, and sedimentary processes and chemical transformations in the coastal environment), and other NSF divisions.

Recommendation: EAR should take the lead within NSF in devising a long-term strategy for funding research on the Critical Zone.

The study of Critical Zone processes would also benefit from partnerships between EAR and other federal agencies. NSF's focus on research initiated by external proposal submission would complement the more directed research on coastal zone processes carried out by NOAA (characterization and assessment of coastal change), the Federal Emergency Management Agency (coastal erosion and flood hazard), and USGS (environmental quality of coastal areas) (see "Related Federal Research Programs"). Similarly, soil research sponsored by USDA (soil properties and carbon sequestration), EPA (soil contaminants and pathways), the Department of Defense (DOD) (soil erosion, compaction, and trafficability), DOE (bioremediation and the carbon cycle), NASA (response of terrestrial life to conditions in space), and a host of private foundations is typically directed toward practical questions of agriculture, environmental quality, and land management. The understanding generated from these applied studies, as well as from basic research funded by EAR, will provide a key contribution to many outstanding Earth science problems.

MAJOR INITIATIVES

Previous EAR initiatives have funded major new tools for Earth observations, as well as new mechanisms for multidisciplinary research (Appendix A). One outstanding example is the program in observational seismology run by the Incorporated Research Institutions for Seismology (IRIS), which now comprises 100 members, primarily U.S. universities and research organizations, and manages an annual budget of about $11 million.[8] IRIS provides the instrumentation and infrastructure for gathering and disseminating seismological

[8]The IRIS consortium, begun in 1984, has been responsible for the deployment of the Global Seismic Network, a worldwide distribution of 120 permanent, high-performance seismic stations sponsored by NSF in cooperative agreements with USGS, DOD, and the Program for Array Seismic Studies of the Continental Lithosphere.

data to the entire Earth science community. These data are critical to the study of the continents and deep interior, two major areas of scientific opportunity discussed in Chapter 2. A second example is the Continental Dynamics (CD) Program, begun in 1984 in response to an NRC report[9] and now an EAR core program with an annual budget of about $9 million. The CD program funds multidisciplinary research that focuses on an improved understanding of the processes governing the origin, structure, composition, and dynamical evolution of the continents; it is thus complementary to the facility-oriented IRIS program.[10]

In the following section, two major research initiatives are considered: (1) EarthScope, a facility-oriented program for observing the structure and active deformation of the North American continent, and (2) an Earth Science Natural Laboratory program (ESNL), which would support long-term, multidisciplinary observatories of terrestrial processes in specific field areas. Funding for the first stage of EarthScope has been proposed in the President's 2001 budget request,[11] whereas the ESNL program is still in the conceptual stage.

EarthScope

EarthScope is an NSF initiative to build a distributed, multipurpose set of instruments and observatories that will substantially enhance the capabilities of Earth scientists to investigate the following research topics:

- Earthquakes and seismic hazards
- Magmatic systems and volcanic hazards
- Active tectonics
- Fluids in the crust

---

[9]*Opportunities for Research in the Geological Sciences*, National Academy Press, Washington, D.C., 95 pp., 1983.

[10]The CD program is particularly oriented toward projects whose scope and complexity require a cooperative or multi-institutional approach and multiyear planning and execution. The program funds only relatively large projects that do not fit easily within other EAR programs and that have broad support within the Earth science community. The program also funds research as part of the interagency Continental Scientific Drilling and Exploration Program.

[11]The request for EarthScope in the President's FY 2001 budget is $17.4 million in FY 2001, $28.5 million in FY 2002, $15.7 million in FY 2003, and $13.2 million in FY 2004. The total request under the NSF's Major Research Equipment program is $74.8 million, which covers only the acquisition, construction, and deployment aspects of EarthScope-Phase I (USArray, the San Andreas Fault Observatory at Depth).

- Continental structure and evolution
- Geodynamics of the mantle and core

As currently conceived, EarthScope will comprise four programmatic elements, each centered around a major new observational effort: (1) *USArray*, for high-resolution seismological imaging of the structure of the continental crust and upper mantle beneath the coterminous United States, Alaska, and adjacent regions; (2) *San Andreas Fault Observatory at Depth* (SAFOD), for probing and monitoring the San Andreas Fault at seismogenic depths; (3) *Plate Boundary Observatory* (PBO), for measuring deformations of the western United States using strainmeters and ultraprecise geodesy of the Global Positioning System; and (4) *Interferometric Synthetic Aperture Radar (InSAR) Initiative*, for using satellite-based InSAR to map surface deformations, especially the deformation fields associated with active faults and volcanoes. Box 2.2 contains a more complete description of these programs.

EarthScope will contribute substantially to understanding the structure, evolution, and active deformation of the continents and the attendant earthquake and volcanic hazards. It will also improve seismic images of the deep interior and furnish important new data on basic geodynamic processes. The committee has been impressed by the broad disciplinary representation and the diverse grass-roots efforts that have contributed to the formulation of the EarthScope initiative; these have included a series of open, well-attended workshops and a number of symposia at national scientific meetings, as well as extensive discussions among leaders in the relevant disciplines to translate the input material into a viable science plan.

Finding: EarthScope will address major science problems related to the continents and deep interior identified in this report. The scientific vision and goals of EarthScope are well articulated and have been developed with a high degree of community involvement. The committee strongly endorses all four components of the EarthScope initiative.

To be successful, the major observational elements of EarthScope will have to be backed by strong disciplinary programs to interpret the data within a larger scientific context. Understanding the structure and evolution of the continents, for example, will require a broad spectrum of geological, geochemical, and geophysical studies, including targeted field work, to interpret the seismic structure imaged by USArray. Such studies should not be confined just to the United States, because many analogous and better-exposed structures in other parts of the world may lead to a deeper and more general understanding. Offshore investigations are required to study effectively even the North

American continent. Similarly, the greatly enhanced observations of the plate boundary deformation field provided by PBO should be complemented by substantial theoretical and experimental modeling efforts, as well as field investigations of active deformation in various settings. In particular, a major paleoseismology effort will be needed to extend the temporal range of earthquake observations into the geological past.

Many of these problems can, with adequate new funding, be addressed through existing EAR programs, including the CD core program, the interagency National Earthquake Hazards Reduction Program (NEHRP), and two special emphasis areas—Active Tectonics (AT) and Cooperative Studies of the Earth's Deep Interior (CSEDI).[12]

Finding: Existing programmatic elements within EAR offer the mechanisms to support the basic science required for a successful EarthScope initiative, but only if funding is adequately augmented for basic disciplinary and multidisciplinary research.

NSF activities under EarthScope will couple to efforts in other NSF divisions and federal agencies. The offshore component of EarthScope will require the deployment of temporary arrays of ocean-bottom seismometers (Figure 2.15), requiring the involvement of OCE, which currently funds these facilities. In turn, data from the offshore components of USArray and PBO will contribute to the Continental Margins Research (MARGINS) program in OCE. Research on earthquake hazards will support NSF's mission in NEHRP, including earthquake-related programs in the Division of Civil and Mechanical Systems in the Directorate of Engineering. All four components of EarthScope will contribute substantially to efforts by the USGS to assess earthquake and volcanic hazards, creating excellent opportunities for interagency cooperation. The InSAR component, which requires new satellite-based observing systems, should be the basis for a substantial cooperative program between NSF and NASA.

Natural Laboratories

Some of the processes operating in the solid Earth can be isolated and studied in the laboratory under controlled conditions, but this approach is

[12]The committee notes that the excellent, community-generated science plans that led to the AT and CSEDI programs anticipated essentially all of the major scientific goals of the EarthScope initiative.

limited by the feasible dimensions of laboratory apparatus, usually a few meters or less, as well as the duration of the experiments. Moreover, it is often impossible to reproduce in the laboratory the range of interactions among processes required to emulate natural phenomena. For these reasons, the study of terrestrial processes almost always requires extensive field observations. One especially effective strategy for dealing with natural complex systems is to focus observations on carefully chosen areas—natural laboratories—where representative behaviors can be investigated in appropriate context and detail and with the appropriate complement of expertise and instrumentation.

Designating specific areas for special scrutiny has several advantages. It facilitates the coordination of activities across multiple groups of investigators, encouraging the types of multidisciplinary studies that are often essential to understanding complex processes and system behaviors. It also provides a long-term basis for capitalizing on field-based research. If the investigations are well directed and the data properly analyzed and archived, then the return on previous research investments can be compounded as more data are collected. Each observational study within the natural laboratory adds to the database, improving the context for future work. This coordinated, multidisciplinary approach is especially desirable when field operations are logistically complicated and expensive, as in the collection of spatially dense data sets and the monitoring of phenomena over extended time intervals. Synoptic studies of natural laboratories furnish an important observational base for developing theoretical and numerical models of complex natural systems, and they yield the essential data by which these models are ultimately validated. They also provide the facilities for involving students and teachers in participating in field-based research (see "Education" below).

The establishment of natural laboratories has become commonplace in programmatic studies of the seafloor sponsored by NSF's Ocean Sciences Division, where coordinated, multidisciplinary research in specified regions has proven to be an effective strategy for studies of seafloor processes and systems and has become essential to the efficient use of ships and other expensive oceanographic facilities.[13] The natural laboratory concept is also the basis for NSF's Long Term Ecological Research (LTER) Network,

[13]An early (1974) example was Project FAMOUS (French-American Mid-Ocean Undersea Study), which coordinated an extensive program to make the first direct observations of seafloor spreading on a segment of the Mid-Atlantic Ridge. OCE's RIDGE program is developing a set of seafloor observatories at various points on the global spreading system, beginning with sites on the Juan de Fuca Ridge, and the MARGINS science plan calls for a concentration on a set of "focus study areas" targeted for intensive, multidisciplinary programs of research that can exploit the synergy among field experiments, numerical simulations, and laboratory analyses.

established in 1980 for investigating ecological processes operating over extended periods (months to centuries) at a variety of spatial scales (from 10 m to continental).[14]

The research areas discussed in Chapter 2 illustrate the potential utility of natural laboratories in the context of Earth science. For instance, the USGS maintains a long-term, multidisciplinary program for the study of earthquake processes on the San Andreas Fault at Parkfield, California, located at the transition between the creeping and locked sections of the fault. Special arrays of surface and borehole instrumentation have furnished insights into seismogenic processes at scales much smaller than typical seismological investigations. Owing to the enhanced understanding of earthquake processes achieved through these observations, the Parkfield natural laboratory has been chosen as the site for the SAFOD component of the EarthScope initiative, which will use deep drilling to conduct in situ investigations of the San Andreas Fault zone at seismogenic depths of 3 to 4 km. On a somewhat larger scale, Southern California has been used as a natural laboratory for earthquake studies by the Southern California Earthquake Center, a consortium of eight universities jointly funded by EAR, the USGS, and NSF's Science and Technology Centers Program. The Office of Naval Research sponsors modeling, experimentation, and multiyear field work at two continental sites with the object of linking process studies with studies of Holocene deposition patterns. Some sites proposed for critical facilities, such as the Yucca Mountain Nuclear Waste Repository and the Ward Valley facility for radioactive waste, have become de facto natural laboratories by virtue of the comprehensive investigations mandated by environmental and hazard-vulnerability concerns. Extensive field work at these and other DOE sites, which includes the mapping of hydrological and chemical fluxes in all three spatial dimensions on time scales ranging from hours to millennia, is producing a much more comprehensive understanding of the geochemical processes within the Critical Zone.[15]

The scientific advances made in the study of these natural laboratories illustrate the potential for multidisciplinary research of near-surface processes

[14]The network promotes synthesis and comparative research across sites and ecosystems, as well as among other related national and international programs. Several federal agencies cosponsor LTER activities with NSF, and 17 other countries have formal LTER programs. Each of the 21 current sites has a data manager and principal investigator. They are funded and reviewed separately on a six-year cycle, and the entire network is reviewed every five years. Recent LTER awards range up to $4.2 million, with a median of $1.8 million. Projects are multidisciplinary and actively encourage collaborations with other investigators; support for such collaborations comes from the relevant disciplinary programs.

[15]See, for example, *Groundwater at Yucca Mountain: How High Can It Rise?* National Academy Press, Washington, D.C., 231 pp., 1992.

in a site-specific context. Floods and droughts (e.g., water budget, flow paths, and predictive capabilities) and the role of soils and biota in geochemical cycles (e.g., carbon sequestration, geology-climate links, and soil and water quality) exemplify the types of problems that have to be monitored at fixed, well-characterized localities on time scales exceeding standard project durations. Such studies of the Critical Zone would observe how the geologic record is created, thus improving its interpretation.

Five independent workshop reports submitted to the committee[16-20] call for the establishment of natural laboratories to exploit scientific opportunities across a range of problem areas. The needs of the research community in this regard have been recognized by EAR, with some success. Multidisciplinary projects to take advantage of natural laboratories have been sponsored by EAR's Continental Dynamics Program, and other short-term efforts have been sponsored under the auspices of EAR core programs. However, because EAR's current programmatic structure does not allocate specific funds for natural laboratories, it has been difficult to match the investments in establishing such natural laboratories with the long-term resources needed to take full advantage of their availability.

Recommendation: EAR should establish an Earth Science Natural Laboratory Program with the objective of supporting long-term, multidisciplinary research at a number of promising sites within the United States and its territories.

The ESNL program should be proposal driven and open to all EAR problem areas and disciplines. The committee notes that natural laboratories would provide especially effective platforms for multidisciplinary studies of surficial, near-surface, and coastal processes in the Critical Zone, and it would thus be appropriate to place special emphasis on the Critical Zone when selecting ESNL sites. As with LTER sites, special requirements should be put in place to ensure that data collected by the program are properly

---

[16]*A Vision for Geomorphology and Quaternary Science Beyond 2000*, results of a workshop held in Minneapolis, Minnesota, February 6-7, 2000.

[17]*Research Opportunities in Low-Temperature and Environmental Geochemistry*, results of a workshop held in Boston, Massachusetts, June 5, 1999.

[18]*Sedimentary Systems in Space and Time: High Priority NSF Research Initiatives in Sedimentary Geology*, results of a workshop held in Boulder, Colorado, March 27-29, 1999.

[19]*Support for Research in Tectonics at NSF*, White Paper from the Division of Structural Geology and Tectonics, Geological Society of America, July 24, 1998.

[20]*Microscopic to Macroscopic: Opportunities in Mineral and Rock Physics and Chemistry*, results of a workshop held in Scottsdale, Arizona, May 28-30, 1999.

analyzed, archived, and distributed to a wide user community. Cosponsorship by other NSF divisions and other agencies, including state and local government agencies, should be encouraged. Mechanisms should be considered to facilitate undergraduate and graduate field-based education at ESNL sites.

SUPPORT OF MULTIDISCIPLINARY RESEARCH

Many basic problems in Earth science encompass a combination of physical, chemical, and biological processes and are thus intrinsically multidisciplinary. Although EAR has always sponsored multidisciplinary research through its core programs, the success of proposals that cross program boundaries has depended in part on the composition of the review panel and the assertiveness of NSF staff in seeking partial funding from other relevant programs. Consequently, programs specifically designed to have a multidisciplinary focus, such as the Continental Dynamics Program and the fixed-term special emphasis areas (Table A.1), have been particularly effective in funding of multidisciplinary research, for both individual investigators and groups of investigators. The major initiatives discussed in the previous section, if implemented, will establish new mechanisms for sponsoring multidisciplinary research. In addition, the committee suggests that EAR initiate fixed-term programs in two research areas discussed in Chapter 2—microorganisms in the environment and planetary science—which offer particular promise for significantly advancing scientific understanding through multidisciplinary studies.

Microorganisms in the Environment

The recent flood of information from the application of both geochemical and biochemical tools has raised exciting questions about how microorganisms interact with geological and pedological processes. New observations reveal how little is actually known of the richness of microbial influences on the global environment and, conversely, of environmental influences on microbial processes.[21,22] Microbial assemblages reflect existing conditions in natural habitats, and their ability to adapt to pressures resulting from human activities in the Critical Zone will have a direct impact on the quality of soil,

---

[21]*Research Opportunities in Low-Temperature and Environmental Geochemistry*, results of a workshop held in Boston, Massachusetts, June 5, 1999.

[22]*Opportunities in Basic Soil Science Research*, G. Sposito and R.J. Reginato, eds., Soil Science Society of America, Madison, Wisconsin, 129 pp., 1992.

water, and air. The ecology and geochemistry of microorganisms are topics of growing importance within the larger field of geobiology. Although studies of microbial diversity and phylogeny have traditionally benefited from the efforts of biogeochemists and other geoscientists, the last few years have witnessed a dramatic expansion of the interest in this area by a large community of geochemists, soil scientists, and ecologists. These efforts would be strengthened by multidisciplinary approaches that integrate geochemical and microbiological insights. Such efforts promise substantial gains in understanding the following:

- microbial interactions with minerals, mineral surfaces, nanophases and materials, metals, and life-sustaining elements,
- impact of microbial activities on the natural and human-influenced environment over spatial scales ranging from atomic to global, and
- geochemical and phylogenetic records of interactions between environmental change and the evolution, diversity, function, and ecology of microorganisms.

Studies of microbial-environment interactions are currently funded by a number of government agencies. For example, water quality and related engineering issues are supported through EPA, USDA, and NSF, whereas NSF's Life in Extreme Environments and NASA's Astrobiology programs fund studies focused on life in unusual habitats. Yet, this support is too narrow and targeted to establish a robust basis for collaborations among the various disciplinary communities. The tools and insights that have recently become available furnish an unprecedented opportunity to revolutionize the understanding of microbial life in geologic environments and the roles that microorganisms play as geological, pedological, and geochemical agents.

Recommendation: EAR should seek new resources to promote integrative studies of the way in which microorganisms interact with the Earth's surface environment, including present and past relationships between geological processes and the evolution and ecology of microbial life.

The Environmental Geochemistry and Biogeochemistry special emphasis area, initiated with EAR participation in 1994, sponsored a wide array of research within environmental science and engineering, including studies of microbial activity in an environmental context. The research proposed here would build on the success of EGB, with a sharpened focus on microbial agents in the environment. The committee notes that support for the study of microbial-mineral interactions could be broadened through the new interagency effort on nanotechnology, which recognizes the importance of microbial interactions with minerals and surfaces.

Research on microorganisms in the environment would serve as an important link between the emerging fields of geobiology and Earth materials and would be a central component of future studies of the Critical Zone. It would be strengthened by increased support for natural and mobile laboratories (see "Instrumentation and Facilities" below), as well as enhanced access to geochemical tools though centers and other facilities. It would benefit from EAR and Education and Human Resources (EHR) support for cross-disciplinary training of students and professionals, discussed below under "Education." There may be significant opportunities for collaborating with other NSF directorates (e.g., Biological Sciences Directorate), as well as with other federal agencies (e.g., DOE, USDA, NIH) in this domain, as outlined in the committee's previous discussion of potential agency partnerships in "Geobiology."

Planetary Science

International exploration of the solar system and extrasolar planets, along with a number of proposed sample return missions, will provide a wealth of new data and materials with which to address basic questions about the origin and evolution of the Earth and planets and, possibly, of life. However, present funding structures and practices are ill-suited to capitalize on the excitement and opportunities these new data and materials will provide. NASA, quite rightly, places a great deal of emphasis on planetary science investigations that are directed primarily toward the design, data collection, and interpretation of results from specific spacecraft missions. However, funding for investigator-driven basic research is becoming increasingly inadequate to support the large and diverse research communities that should be engaged in the new era for planetary science. Although EAR programs consider proposals for basic research in planetary sciences, very little EAR funding is actually invested in such research. NSF-Astronomy sponsors telescopic investigations of solar and extrasolar planets but is not a logical home for the complementary types of planetary research discussed in Chapter 2. Without support for broad-based planetary science in NSF, there is the danger that important opportunities in this field will be missed as new data sets and extraterrestrial samples accumulate over the next 10 years.

Recommendation: To promote increased interactions between the Earth and planetary science research communities and to exploit the basic research opportunities arising in the study of

solar and extrasolar planets, EAR should initiate a cooperative effort with NASA and NSF-Astronomy in planetary science.

The scope of a multiagency cooperative program in planetary science would likely include broad-based investigations of the fundamental processes that produce and modify planets using an array of geophysical, geological, geochemical, and modeling approaches. EAR is thus a logical leader of the proposed collaborative endeavor.

To ensure the vitality of this effort, EAR should seek new funds to establish a program in planetary sciences. The program should be sufficiently large to support a robust and healthy research community, including individual investigators, teams of investigators, and the instrumentation critical for ground-based planetary science investigations. Such a program would have natural links to other high-priority research efforts within and beyond NSF, including the intensive investigation of Mars and LExEn. It may ultimately be appropriate to establish planetary sciences as a new core program within EAR.

INSTRUMENTATION AND FACILITIES

The EAR Instrumentation and Facilities (I&F) Program has played a crucial role in the development of many new research tools, ranging from major facilities, such as synchrotron beamlines, accelerator mass spectrometers, and seismic arrays and networks, to the development and dissemination of analytical, computational, and information technologies in individual laboratories. Sustained support of this type is essential for high-quality experimental research. In the coming decade, this program will be subjected to the multiple stresses of rising equipment, operation, and maintenance costs. To take advantage of new technologies, EAR will no doubt have to expand the resources devoted to major research facilities. An unprioritized list of areas with a growing need for instrumentation and facilities support includes the following:

- *Neutron-scattering facilities for the study of minerals, rocks, soils, and other planetary materials*: New intense sources coupled with state-of-the-art detectors will allow dynamical (inelastic scattering) as well as structural (diffraction) measurements on large (milliliter to liter) samples, including at high or low pressure and temperature. Determining sites of hydrogen in minerals, measurements of phonon density of states or magnetic properties on small samples, specific surface properties, chemisorbed speciation, characterization of large textured rock samples, analysis of aperiodic and structurally complex Earth materials (e.g., liquids, glasses, soil components), including biomineral composites, order-disorder, and microcrystallography of planetary materials, are among the important applications that are especially suited to this approach. The geosciences community should provide leadership in the development of these technologies, as it has successfully done with synchrotron facilities.

- *"Smart" synchrotron beamlines, fully instrumented for multiple probes simultaneously characterizing a sample with high spatial resolution*: For example, an X-ray/infrared beamline could be combined with sophisticated optical, magnetic, and laser instrumentation to probe processes on many different length and time scales (e.g., in situ determination of acoustic velocities and crystal structures of samples at the high-pressure and temperature conditions of the Earth's deep interior).
- *New laser-based experiment and analysis technologies*: These include (1) laser-driven shock-wave methods for achieving 1-10 TPa pressures relevant to reproducing giant-planetary and small-star (or brown-dwarf) interiors, and (2) fourth-generation synchrotron beams, whether based on "tabletop" lasers or large free-electron laser systems.
- *Geochemical facilities and instrumentation*: The growing need for geochemical analyses with higher precision, spatial resolution, and detection limits necessitates the development of new instrumentation. These advances will be needed to more fully characterize both terrestrial and extraterrestrial samples in the coming decade. The instrumentation will include multiple-collector ion probes, inductively coupled plasma mass spectrometers, and advanced electron microscopes. In addition, studies of the Critical Zone will require far more detailed characterization of organic substances than presently available, using techniques such as nuclear magnetic resonance, electron paramagnetic resonance, and various types of mass spectrometry of organic molecules.
- *Improved access to geochronometry*: There is an increasing need for routine access to rapid, high-precision dating, which is particularly acute in fields requiring ages, especially radiocarbon ages, determined by accelerator mass spectrometry (AMS). The capacity of existing AMS facilities is inadequate to meet current demands. Improvements should be made to the throughput at current multiuser dating facilities, and the construction of a new-generation AMS facility should be considered.
- *Mobile instrumentation for ground-based remote sensing of key hydrologic variables* (e.g., atmospheric moisture content, wind speeds, and cloud characteristics): Relevant instruments include dual-polarization Doppler radar, lidar, and Doppler sound detection and ranging wind profiler. Enhanced ground-penetrating radar, capable of more fully characterizing the subsurface, and fixed ground-based hydrologic instrument clusters will also be required.
- *Microbiological instrumentation and facilities, including mobile laboratories: Facilities are needed for culturing organisms under a variety of conditions*, including anaerobic chambers, equipment for amplifying and

sequencing genetic materials, and computers for analyzing large databases of genomic information, as well as for analyses of surface and mineralogical characteristics by high-resolution techniques such as atomic force microscopy. Open-path CO_2, organic, and inorganic carbon analyzers will be needed for studies of carbon sequestration in soils. Field equipment to preserve the redox state of soils and sediments sampled from aqueous environments will be needed for geochemical interpretation of metal toxicity, fate, and transport.

- *Computers and access to computing facilities*: Innovations in Internet connectivity, multimedia information processing, digital libraries, and visualization techniques are needed to expedite the collection, dissemination, and processing of heterogeneous streams of data from an expanding array of observatories. Improvements in modeling will require high-speed access to distributed computing facilities, algorithms that utilize computational grids, and structures for developing community models.

A second stress arising within the I&F program concerns the operation and maintenance of instrumentation. An example relevant to large-scale facilities is the maintenance and modernization of the Global Seismic Network (GSN). The GSN is critical for many aspects of Earth science, but support for these purposes has historically been problematic. A long-term strategy is an important issue for NSF and USGS, particularly since a substantial shortfall in funding (approximately $3 million) for the USGS component already exists.

Although the I&F program has been very effective in allocating equipment to individual principal investigators, it places a lower priority on funding the operation and maintenance of equipment *after* it is purchased. As currently implemented, the program gives technician support to individual laboratories for a maximum of five years. Full technician costs are rarely funded through the disciplinary programs after this period. As a result, EAR principal investigators have been forced to seek ongoing support from other sources, such as contract work or institutional discretionary funds, with mixed success. The lack of continuity in laboratory operation and technical staffing is a growing problem. Academic researchers find it increasingly difficult to keep expensive equipment operating efficiently and to hire and retain highly qualified technicians and engineers, who often seek greater security and higher pay in positions outside universities.

As equipment and personnel costs rise, researchers are turning toward multiuser instrumental facilities. Community-based facilities can increase the efficiency and reduce the burden of maintaining expensive equipment at many universities, and they form a basis for establishing community priorities. Indeed, several workshop reports emphasized the desirability of this approach.[23-25] These advantages are exemplified by the successful University NAVSTAR Consortium (UNAVCO) and the Program for the Array Seismic Studies of the Continental Lithosphere (PASSCAL). Multiuser facilities tend to be effective when the instrumentation is standardized and the data analysis is relatively routine and can be done in high volume.

> **Recommendation: EAR should seek more resources to support the growing need for new instrumentation, multiuser analytical facilities, and long-term observatories and for ongoing support of existing equipment.**

Communal facilities should have the resources to provide state-of-the-art equipment, long-term technician support, and support for visitors and workshops. A regular mechanism should be established to evaluate the success of multiuser facilities in meeting the demands of a recognized user community.

Although multiuser analytical facilities are a cost-effective means to support the expensive instrumentation needed by a broad investigator community, they are often less effective for developing new technologies or tailoring individual analyses to specific requirements. This has been especially (but not exclusively) conspicuous in geochemistry, where much of the field moves forward through a combination of diverse developments within individual laboratories. Similarly, multiuser facilities are inappropriate when the analysis of one sample may adversely affect the analysis of another (e.g., because of blank levels).

The current approach to this problem has been to support a large number of research laboratories. Young investigators are encouraged to establish independent laboratories when they are hired into faculty positions. After a period of initial institutional support, investigators are generally expected to cover operating costs from outside (usually federal) sources, placing an ever-increasing burden on NSF. Although this approach has been extremely successful in developing a robust investigator base, the resulting growth will be difficult to sustain in the long run. The number of EAR-sponsored laboratories must clearly be sufficient to support a vibrant research community, but it is becoming necessary to explore the trade-off between this number and the level of support available to individual laboratories.

---

[23] *A Vision for Geomorphology and Quaternary Science Beyond 2000*, results of a workshop held in Minneapolis, Minnesota, February 6-7, 2000.

[24] *Research Opportunities in Low-Temperature and Environmental Geochemistry*, results of a workshop held in Boston, Massachusetts, June 5, 1999.

[25] *Sedimentary Systems in Space and Time: High Priority NSF Research Initiatives in Sedimentary Geology*, results of a workshop held in Boulder, Colorado, March 27-29, 1999.

Recommendation: The I&F program should encourage its user communities to identify research priorities and develop a consensus regarding how many laboratories are needed and how their operational costs should be apportioned among the EAR core programs, the I&F program, and participating academic institutions.

After initial instrument commissioning, it may be appropriate to encourage technician support as a routine cost of ongoing research projects rather than through specific I&F grants for technician support. The issue is complex because it involves a deeply rooted aspect of the culture—individual laboratories—that many investigators view as the heart of innovation in their field. It also has an obvious bearing on the continued flow of vigorous young investigators into these fields.

EDUCATION

The debate on many of the social issues facing our nation and the world benefits from increased scientific literacy in the general population. In turn, science benefits from a population that understands the nature of scientific inquiry and its value to society. As described in Chapter 1, the most pressing societal issues (e.g., resource sustainability, mitigating natural hazards, managing the environment) have an Earth science component. Thus, knowledge of the Earth sciences must be part of the background of every informed person.

NSF has reaffirmed science education, along with basic research, as an agency priority. Indeed, NSF funds research in part because it results in the best possible environment for higher-level education. The research-based education approach as exemplified by U.S. research universities has been so successful that it is being used as a model for restructuring universities in other countries, particularly in Europe and Japan. Research-based education is funded mainly through NSF-wide initiatives and the science directorates. The Directorate for Education and Human Resources, on the other hand, focuses on the science of education—teaching and learning.

A 1996 workshop[26] challenged the Geosciences Directorate to promote vigorously educational activities within its research program and to enhance its partnership with EHR, beginning with helping geoscientists understand

[26]*Geoscience Education: A Recommended Strategy*, results of a workshop held in Arlington, Virginia, August 29-30, 1996. The report outlines a strategy for improving outreach to teachers and other communities, enhancing university-level training with emphasis on links to nonresearch needs, and facilitating the educational value of ongoing programs ranging from research consortia to undergraduate institutions.

and submit proposals to EHR programs. The committee strongly endorses such efforts, particularly programs that support the involvement of undergraduate and secondary students in basic research projects and encourage the broadest participation in Earth science through Internet accessibility and participation—including digital libraries.

Within EAR, there are many opportunities for blending education with basic research. The challenge for EAR is to build flexible programs that (1) are appropriate to the scale and topic of different research projects; (2) encourage the dissemination of research results to a wide audience, ranging from colleagues to the general public; and (3) build in support for education that complements, rather than competes with, support for basic science.

Training in the Earth Sciences

Earth science training is becoming increasingly demanding. Not only must Earth scientists keep pace with developments in physics, chemistry, biology, and engineering, they must also be cognizant of the social and economic influences of their work. Indeed, the attempts by humans to manage the terrestrial environment on a planetary scale raises many ethical, political, and philosophical issues.[27] Thus, a key challenge for educators is to develop pre-college and undergraduate curricula in Earth science that encompass a wide variety of knowledge and approaches. This is particularly important because many Earth science graduates go on to work in unrelated fields. According to NSF's National Survey of College Graduates, only 26% of the recipients of B.S. degrees in Earth science are employed in the same science field, and nearly 60% have nonscience occupations, mainly in industry.[28] A majority of M.S. recipients are working in the Earth sciences (50%) or a related field (20%), although many combine science with other tasks, such as management, sales, computer applications, professional service, and teaching. Surprisingly, Ph.D. recipients are almost as likely to work in a related science discipline (37%) as in Earth science (46%). Half of the Ph.D. recipients are employed in the education sector; the remainder work in business-industry-nonprofit (30%) or government (20%).

The current funding structure at NSF requires most training of undergraduate and graduate students to come through focused research projects,

---

[27]*Research Priorities in the Geosciences: Philosophical Perspectives*, Results of a workshop held in Boston, Massachusetts, June 5, 1999.

[28]In the survey, "Earth scientists" includes atmospheric, Earth, and ocean scientists. See http://srsstats.sbe.nsf.gov.

typically individual investigator grants. Although such training teaches students important disciplinary skills, it can lack the cross-disciplinary components that allow students to branch into new fields and gain the knowledge needed to pursue a broad range of career opportunities. Better integration of education and research in a multidisciplinary framework is needed to help students take advantage of new research directions and employment options. Grants and fellowships that offer this flexibility, such as those sponsored by NSF's Integrative Graduate Education and Research Training (IGERT) Program,[29] are key to attracting and retaining the best students. A plausible strategy for encouraging broad-based training is for EAR to establish a program of research training grants, either on its own or with other science divisions, to provide undergraduates and graduate students with access to alternative research environments.[30] Training grants available through other NSF divisions could serve as models for an EAR program.

Recommendation. EAR should institute training grants and expand its fellowship program to facilitate broad-based education for undergraduate and graduate students in the Earth sciences.

Earth scientists who have just begun their careers (i.e., postdoctoral researchers), as well as those who are already established (i.e., professors), need help in bridging their research to other disciplines. For example, researchers with adequate backgrounds in both biology and the Earth sciences are needed to advance the field of geobiology. Similar training across disciplinary lines is required for study of the atmosphere-hydrosphere-lithosphere-pedosphere interactions that govern the long-term behavior of the Earth's climate, the problems of comparative planetology, and many of the other science opportunities discussed in this report. Plausible mechanisms for these purposes involve the postdoctoral training of young scientists and sabbatical-leave opportunities for established academic scientists.

Recommendation. EAR should establish postdoctoral and sabbatical-leave training programs to facilitate development of the cross-disciplinary expertise needed to exploit research oppor-

---

[29]The NSF-wide IGERT program was initiated in 1997 to encourage the development of multidisciplinary curricula in doctoral-level education; see http://www.nsf.gov/igert.

[30]As envisaged in the workshop report *Geoscience Education*, such training programs would prepare students for non-academic jobs by offering internships with industry, museums, nonprofit organizations, or government agencies, or for nontraditional research positions by offering fellowships that span multiple programs, disciplines, or institutions.

tunities in geobiology, climate science, and other interdisciplinary fields.

Broadening Involvement in Active Research

Many investigators are interested in involving students and teachers in their research, and most are enthusiastic about sharing their results with the broader community. The Research Experience for Undergraduates (REU) Program and the Science and Technology Centers help to involve undergraduates in research projects led by individual investigators and consortia. This approach should be expanded to allow greater involvement of others, such as secondary school students and K-12 teachers. The success of the REU program demonstrates that research by undergraduates can contribute to basic science, particularly in field-based studies. Creative projects benefit participants and carry science to the broader community in ways that conventional outreach programs might not.

Field-Focused Opportunities

Field work is a fundamental and distinctive aspect of the Earth sciences. It provides the basis for understanding a variety of Earth processes and for validating and calibrating model, laboratory, and remote-sensing results. In addition, field work stimulates students from many backgrounds and helps them develop an appreciation for basic and applied problems in Earth science. Throughout the United States, field sites in many settings—urban, rural, and wilderness—are readily accessible, and many types of field projects can be done at relatively low cost. Thus, field work producing high-quality research data can be sponsored by junior colleges and undergraduate institutions, as well as by research universities.

Field work is currently funded under the relevant core research programs through the normal competitive grant process. However, this route is not generally suitable for student-oriented field projects, even for graduate students. In theory, field work for students could be funded through the REU and Research in Undergraduate Institutions programs, although neither explicitly includes a field component. EAR could also seek other mechanisms for supplementing this small-science endeavor. Field programs at other agencies, such as the USGS and USDA summer intern programs or the Educational Component of the National Mapping Program, could serve as alternative models for establishing and funding graduate and undergraduate field programs. The Earth Science Natural Laboratories recommended by the committee would provide superb settings, as well as an expanded infrastructure for staging field-oriented educational programs for students and teachers.

Recommendation. EAR should take advantage of the broad appeal of field work, its modest cost, and its ability to capture the enthusiasm and research effort across a wide range of institutions by providing sufficient funding for graduate and undergraduate field work.

Education is intrinsic to all basic research, but there is no one-size-fits-all formula for enhancing the educational component. Uniform educational results should not be expected. By utilizing a variety of approaches, EAR will gain the flexibility it needs to create new educational opportunities within the context of the basic research mission and to take advantage of the rapid changes caused by the information revolution.

PARTNERSHIPS IN EARTH SCIENCE

The committee has highlighted ways in which EAR might participate in a number of existing interagency programs and initiate new programmatic partnerships to strengthen Earth science and realize the opportunities discussed in this report. Partnerships among federal agencies have become increasingly important mechanisms for organizing and sustaining large scientific efforts on problems of national interest.[31] Such partnerships have a fourfold rationale: (1) to foster the development of multidisciplinary communities needed to address the high-level problems of complex systems; (2) to translate the results of basic research into practical applications; (3) to leverage the limited resources available to individual programs, including equipment and facilities; and (4) to coordinate research across agency programs, thereby promoting synergies and reducing the duplication of effort. These objectives are especially compelling in Earth science, where high-priority national needs require the coordination of basic and applied research over a range of difficult system-level problems. Realizing the benefits of interagency collaborations can be problematic,

---

[31]A prominent example in geoscience is the U.S. Global Change Research Program, which coordinates research on global environmental changes across all federal agencies, interfaces U.S. efforts with the Intergovernmental Panel on Climate Change and other international assessments, and reports annually to the President and Congress on research results and their implications for federal policies; see http://www.usgcrp.gov.

however, owing to structural problems among the participating agencies and scientific communities, as well as practical management issues.[32]

Partnerships Within NSF

As noted in this chapter and described in Appendix A, programs potentially suitable for funding certain aspects of the research opportunities already exist elsewhere within NSF, particularly in the Biological Sciences Directorate (geobiology and microorganisms in the environment), Ocean Sciences Division (geobiology, shoreline aspects of the Critical Zone, offshore components of EarthScope), Astronomical Science Division (planetary science), Materials Research Division (Earth and planetary materials), and Atmospheric Sciences Division (Critical Zone). In some cases, individual Earth scientists will be able to take advantage of these existing programs, but individual successes in programs external to EAR are not sufficient to mount the strong Earth science efforts envisaged in this report. The committee observes that interdivisional partnerships are most effective when backed by well-defined scientific communities within each of the participating NSF divisions, from which proposals can be solicited and membership drawn for topical workshops and proposal review panels. If enacted, the committee's recommendations regarding core programs will help to franchise several disciplinary communities within EAR, including geobiology, Earth and planetary materials, soil science, and coastal zone studies. With improved core support, these communities will be better organized to participate in interdivisional programs.

Disciplinary organization within the EAR framework, including the identification of active program managers with responsibilities to specific research communities, is particularly important for effective participation in the NSF-wide crosscutting and interdisciplinary programs. EAR is formally associated with three of these interdirectorate initiatives: ESH, EGB, and LExEn (see Box A.2). However, as noted elsewhere in this report, there is great potential for significant participation by Earth scientists in at least three other initiatives: the interagency National Nanotechnology Initiative, Biocomplexity in the Environment, and Information Technology for the Twenty-First Century (IT^2).[33] The

[32]Some of the generic problems associated with federal support of interdisciplinary research are summarized by N. Metzger and R.N. Zare (*Science*, v. 283, p. 642-643, 1999).

[33]The multiagency IT^2 initiative is aimed at pushing the envelope for research and development in information technology, including software, information technology education and work force, human-computer interface, and information management, see http://www.ccic.gov/it2/.

establishment of core programs in Earth and planetary materials and geobiology will make EAR a more effective federal partner in both the nanotechnology and biocomplexity initiatives, respectively.

Although the programmatic mechanisms are perhaps less straightforward, it is clear to the committee that the participation of Earth scientists in federal information technology programs has to be improved. For example, although a series of proposals involving EAR-based scientists has been submitted to the Knowledge and Distributed Intelligence (KDI) program, their success rate in the two competitions was zero.[34] This is peculiar given that the three principal foci of the KDI program—knowledge networking, learning and intelligent systems, and new computational challenges—are clearly applicable to a wide range of Earth science problems. These failures suggest that EAR should be more aggressive in fostering substantial collaborations between Earth scientists and the information technology research community.

In some cases, the disciplinary strength and community organization within EAR are already sufficient to engage other NSF divisions, and the need is primarily for NSF managers to provide a structure for interdivisional collaborations. One obvious opportunity is the cooperation between EAR and OCE needed to manage the offshore components of the EarthScope initiative. EarthScope will also provide an expanded basis for interaction between EAR and the MARGINS program. In addition, there are excellent opportunities for strengthening links between EAR and the Division of Civil and Mechanical Systems (CMS) in NSF's Directorate for Engineering. The CMS division supports "research that will increase geotechnical knowledge for foundations, slopes, excavations, and other geostructures, including soil and rock improvement technologies and reinforcement systems; constitutive modeling and verification in geomechanics; remediation and containment of geoenvironmental contamination; transferability of laboratory results to field scale; and nondestructive and in situ evaluation."[35] Many of these topics are relevant to the National Earthquake Hazards Reduction Program and the Network for Earthquake Engineering Simulation, which provides a basis for cooperation between EAR, CMS, and other government agencies.

---

[34]For a listing of KDI proposals that have received awards to date, see http://www.ehr.nsf.gov/kdi/default.htm.

[35]From the programmatic description at http://www.eng.nsf.gov/CMS/CGS/cgs.htm.

Partnerships between EAR and Other Agencies

In Chapter 1, the case was made that basic research in Earth science is relevant to societal needs in five specific areas of application: (1) discovery, use and conservation of natural resources; (2) characterization and mitigation of natural hazards; (3) geotechnical and materials engineering for commercial and infrastructure development; (4) stewardship of the environment; and (5) terrestrial surveillance for global security and national defense. Although the role of NSF is to fund basic research, it is the mission of other federal agencies to apply this research to national problems. Most federal agencies support a mixture of basic and applied research in areas specifically related to their respective missions, but in many cases the basic research components are not the principal thrust or are narrowly constrained.[36] Even among the few agencies with strong basic research programs, such as the USGS, none rivals EAR in the breadth and depth of the Earth science it sponsors. Therefore, the effective translation of basic research to practical applications requires meaningful collaborations between NSF and mission-oriented agencies, especially when the applications are based on an understanding of the complex natural systems obtained through a broad spectrum of multidisciplinary research.

Disciplinary identification and organization under the NSF framework are a prerequisite for effective interagency collaborations. For example, the committee has already pointed out that the absence of a programmatic home for soil science in EAR has made the evaluation and funding of basic soil science difficult, even though this field is quite relevant to research programs in several NSF divisions, as well as to applied research in other federal agencies, particularly the USDA and USGS. The committee's recommendation for accommodating soil science more formally in a geology core program is aimed in part at franchising soil scientists so that they can more effectively participate in wider initiatives. Indeed, the deep intellectual connections made through fundamental research furnish very effective pathways for broadening communities beyond the narrow specialties of individual researchers and focused research groups. This perspective motivates the committee's optimism that a rich spectrum of collaborations among geobiologists, geochemists, hydrologists, geomorphologists, and soil scientists on problems of the Critical Zone will lead to practical benefits for society. It also illustrates why EAR should take the lead in forging partnerships in Earth science between NSF and mission-oriented federal agencies.

EAR has a long history of partnerships with a number of agencies, primarily USGS, DOE, and NASA, on joint research projects and equipment

[36]N. Metzger and R.N. Zare, *Science*, v. 283, p. 642-643, 1999.

acquisition (see Table A.3). Such partnerships have leveraged the science that EAR is able to support, and there are tangible reasons for expanding them in the future. For example, there must be continuing interaction between EAR and USGS on the Advanced National Seismic System (ANSS) and EarthScope, which are closely related and complementary initiatives.[37] It would therefore be effective and beneficial to present the ANSS and EarthScope projects to Congress as a coordinated budget request. An exceptionally promising example of where close interagency collaboration will be fruitful is in the development of InSAR capabilities for measuring active deformation, another major objective of EarthScope. The satellites capable of InSAR imaging will be flown by NASA, but the interpretation of data requires the integration of InSAR data with field studies and other ground-based data collection efforts, activities that should be sponsored through EAR. Indeed, laboratory and field studies supported by EAR are essential for calibrating, validating, and helping to interpret a wide variety of remote-sensing measurements, including gravity, magnetic, and geodetic observations carried out by NASA and other agencies.

REQUIRED RESOURCES

The committee's recommendations, taken together, lay out a basis for the way in which EAR can respond to major Earth science challenges and opportunities in the next decade. It should be noted that, in developing its recommendations, the committee did not review the existing EAR program or other federal research programs. Rather, it focused on new research areas that could be added to the EAR portfolio. Consequently, the budget estimates given below will have to be evaluated in a broader context that was not possible in this study.

The committee estimates that the new funding needed to implement these recommendations would increase the EAR budget by about two-thirds (Table 3.1). This increase will help to offset the recent decline in federal support of basic Earth science and will substantially strengthen the national effort in this important area of fundamental research.

---

[37]The EarthScope science plan calls for an upgrading of 30 stations of the U.S. National Seismograph Network to the higher-performance standards of the NSF-sponsored GSN, which will benefit USGS in its mission of monitoring earthquakes, while the ANSS deployment plan calls for upgrading regional seismic networks that will assist the EarthScope community in imaging the continental crust and upper mantle at higher resolution.

TABLE 3.1. Estimated Costs of Proposed Programs

Program	New Funds (million dollars per year)
Core Programs:	
Geobiology	7
Earth & Planetary Materials	5
Hydrology	5
Instrumentation and Facilities	10
Major Initiatives:	
EarthScope	10[a]
Natural Laboratories	20[b]
Multidisciplinary Research:	
Microorganisms in the Environment	4
Planetary Science	4
Education	3
Total	68

[a] Exclusive of EarthScope facilities funded through the MRE program.
[b] Includes $5 million for instrumentation and mobile laboratories.

In constructing the budget estimates in Table 3.1, the committee considered the following points:

- The average annual expenditures of the disciplinary core programs in EAR is about $11 million, while EAR contributions to special emphasis areas range from $1 million to $11 million (Appendix A). Programs with total budgets of less than a few million dollars rarely develop thriving constituencies; the committee thus refrained from recommending new programs below this level. The budget figures in Table 3.1 correspond to the annual costs of programs in current dollars.
- A substantial new core program in geobiology ($7 million) is needed to take advantage of exceptional scientific opportunities and to furnish an appropriate focus for EAR participation in the NSF-wide programs in biocomplexity and the environment. In addition, the committee recommends significant funding ($4 million) for a special emphasis area in microorganisms in the environment, which would jump-start the expanded EAR effort in geobiology.

The combination of these two initiatives equals the average budget of the established disciplinary core programs—the minimum level plausible for long-term support of geobiology. The allocation of new funds to geobiology would free resources within the G&P core program for expanded research on the Critical Zone.

- Although the hydrology community is large and well established, the NSF core program in hydrologic sciences is relatively new and expends less than $3 million on investigator-initiated research projects. Correspondingly, the success rate of hydrology proposals (19%) is far below the EAR average (31%). The recommended increment in funding ($5 million) would bring hydrologic sciences in line with other disciplinary core programs and the research opportunities available to this field.
- The community that does basic research on Earth and planetary materials is also well defined and serves the full breadth of the geosciences, though it is smaller than the major Earth science disciplines. The recommended budget in this area ($5 million) would be sufficient to establish a new core program and to fund EAR participation in the National Nanotechnology Initiative.
- Substantial funding ($4 million) is allocated to promote interactions between the Earth science and planetary science communities and to exploit research opportunities arising in the comparative study of solar and extrasolar planets, with a focus on the use of Earth science techniques for the analysis of samples returned from extraterrestrial objects and data on planetary surfaces and interiors.
- EAR is requesting substantial funds from Major Research Equipment (MRE) for the equipment and facility components of the EarthScope initiative. The $10 million in Table 3.1 is for the funding of basic research that uses EarthScope data; it thus excludes the MRE equipment and facility costs. The special emphasis areas in AT and CSEDI, which are currently budgeted at about $1 million each, would be appropriate programs to handle the EarthScope-related increments to the research funding.
- The largest single item in Table 3.1 is for the ESNL program. The committee envisages a program that would expend an average of about $4 million per year on each natural laboratory, with at least five laboratories active at any given time. For comparison, the average cost of LTER sites is about $2 million. However, ESNLs would typically be more expensive because of the high cost of subsurface sampling and imaging. For example, the projected six-year cost of the SAFOD project, a component of EarthScope (see Box 2.2), is $27 million, or $4.5 million per year. A significant fraction of ESNL funding (about one-quarter) would probably be devoted to instrumentation and mobile laboratories and would thus supplement the I&F core program.

- The I&F program has grown to comprise one-quarter of the EAR budget, a level of funding the committee deems necessary to maintain and improve facilities critical for basic research. The recommended level of funding ($10 million), combined with the I&F components of natural laboratories ($5 million), would retain the current balance between the research programs and I&F.
- EAR currently spends about $3 million on education, or about 3% of its annual budget. Given the opportunities for enhancing research-based education and its importance for the future of Earth science, a rough doubling of this budget is plausible. For example, the Geosciences Directorate's contribution to the IGERT program alone was $10 million in FY 1999. The committee takes no position on how these funds would be allocated between EAR-Education and Human Resources and other core programs.

Appendixes

Appendix A

Earth Science Programs

NSF EARTH SCIENCE DIVISION

The Earth Science Division (EAR) of the National Science Foundation (NSF) supports research in the solid-Earth sciences—geology, geochemistry, geophysics—and continental hydrology. EAR is part of the Geosciences Directorate of NSF, along with the Ocean Sciences Division (OCE) and the Atmospheric Sciences Division (ATM).

EAR supports both long-term core programs and fixed-term special emphasis areas. The core programs provide a mechanism for funding unsolicited proposals from individual investigators or small groups of investigators (two to four principal investigators [PIs]). They include disciplinary research programs (geology and paleontology, geophysics, hydrologic sciences, petrology and geochemistry, and tectonics), multidisciplinary research programs (continental dynamics), education and human resources, and instrumentation and facilities. A brief description of the core programs and their FY 1999 budgets is given in Box A.1.

Special emphasis areas are chosen periodically, usually in response to national research initiatives identified by NSF or science advisory committees. They are announced separately from the core programs and have defined goals that are evaluated by special criteria in addition to the standard NSF requirements for scientific excellence. The core programs are "taxed" to pay for these special emphasis areas. EAR is currently supporting two special emphasis areas in the solid-Earth sciences (Active Tectonics [AT] and Cooperative Studies of the Earth's Deep Interior [CSEDI]), along with several other multidivision or NSF-wide programs in research, education, and science and technology centers (see Box A.2, Table A.1). Most of these programs are partially funded by other divisions of NSF or other federal agencies (see "EAR Collaboration with Other Agencies" below).

Box A.1. EAR Core Programs

Geology and Paleontology ($12.3 million). Supports studies of physical, chemical, geological, and biological processes at or near the Earth's surface and the landform, sediments, fossils, low-temperature fluids, and sedimentary rocks they produce.

Geophysics ($17.2 million). Supports research related to the composition, structure, and processes of the Earth's interior.

Hydrologic Sciences ($6.9 million). Supports research dealing with the Earth's hydrologic cycle and the role of water on and near the continental surfaces of the Earth.

Petrology and Geochemistry ($12.4 million). Supports research on igneous, metamorphic, and hydrothermal processes that occur within the Earth and other planetary bodies and on the minerals, rocks, fluids, and ore deposits resulting from these processes.

Tectonics ($8.4 million). Supports research related to understanding the tectonic history of the lithosphere through time.

Continental Dynamics ($8.7 million). Supports multidisciplinary research on the origin, structure, composition, and dynamical evolution of the continents and continental building blocks. The program is oriented toward (1) projects that are not easily funded under the above programs, and (2) projects whose scope and complexity require a multi-institutional approach and multiyear planning and execution.

Education and Human Resources ($3.0 million). Coordinates the division's efforts to improve Earth science education and provides liaison between the Earth science research community and NSF's Directorate for Education and Human Resources.

Instrumentation and Facilities ($26.3 million). Supports the acquisition or upgrade of equipment required for research, the development of new instrumentation and techniques, the operation of multiuser or national facilities, and the funding of research technicians.

Box A.2. EAR Special Emphasis Areas and Cross-Cutting Programs

Active Tectonics. Supports basic research in tectonically active systems of the Earth's continental crust.

Continental Margins Research. Supports research on understanding the interaction of mechanical, chemical, biological, and fluid processes that govern the initiation, evolution, and destruction of continental margins, as well as the accumulation of resources in these regions.

Cooperative Studies of the Earth's Deep Interior. Supports collaborative, interdisciplinary studies of the Earth's interior, including the character and dynamics of the Earth's mantle and core, their influence on the evolution of the Earth, and processes operating within the deep interior.

Earth System History. Supports research on understanding the natural variability of the Earth system (land, ocean, atmosphere), including the forcing mechanisms, interactions, and feedbacks among its components, with annual to millennial resolution.

Environmental Geochemistry and Biogeochemistry. Supports interdisciplinary research on chemical processes that determine the behavior and distribution of inorganic and organic materials in environments near the Earth's surface.

International Continental Drilling Program. Supports drilling, on a global scale, to advance knowledge of the composition, structure, and processes of the Earth's crust.

Life in Extreme Environments. Supports research on microbial life forms and the extreme environments in which they exist, with the goal of detecting and understanding the life forms that may exist beyond Earth.

National Earthquake Hazard Reduction Program. Supports research on earthquake processes as the foundation for applying earthquake hazard reduction measures.

Water and Energy: Atmospheric, Vegetative and Earth Interactions. Supports research on the Earth's hydrologic and energy cycles with the goal of assessing the potential impact of human activities on these cycles and the climate system in general.

Water and Watersheds. Supports systems-oriented research at the watershed scale in a multidisciplinary framework, including physical or engineering, biological, and social sciences.

TABLE A.1. EAR Special Emphasis Areas

Program Title	FY 1999 EAR Portion (million dollars)	Source of Funds
Research Programs		
AT	1.0	EAR/tectonics
CSEDI	1.0	EAR/geophysics
ESH	6.1	EAR/geology and paleontology, EAR/continental dynamics, EAR/instrumentation and facilities, ATM, OCE, OPP, NOAA
EGB	2.3	EAR/hydrologic sciences, ATM, OCE
ICDP	0.7	EAR/continental dynamics, USGS
LExEn	0.3	EAR/geology and paleontology, BIO, ENG, ATM, DMS, OPP, NASA
MARGINS	0.3	EAR/continental dynamics, OCE, ODP
NEHRP	11.3	EAR/continental dynamics, EAR/instrumentation and facilities, EAR/geophysics, ENG, USGS, NIST, FEMA
RUI	not avail.	Individual EAR programs
Water and Watersheds	0.3	EAR/hydrology, EPA, USDA
WEAVE	2.5	EAR/hydrologic sciences, EAR/instrumentation and facilities, ATM, BIO
Science and Technology Centers		
CHiPR	1.8	EAR/instrumentation and facilities, OIA
SCEC	2.5	EAR/geophysics, OIA
Education Programs		
CAREER	0.2	Individual EAR programs, NSF-wide
Postdoctoral Research Fellowships	0.7	EAR/education and human resources, NSF-wide
REU sites	0.6	EAR/education and human resources, NSF-wide
RPG	0.2	EAR/education and human resources, NSF-wide
Other educational activities	0.6	EAR/education and human resources

NOTE: BIO = Biological Sciences Directorate (NSF); CAREER = Faculty Early Career Development; CHiPR = Center for High Pressure Research; DMS = Mathematical and Physical Sciences Directorate (NSF); EGB = Environmental Geochemistry and

Biogeochemistry; ENG = Directorate for Engineering (NSF); EPA = Environmental Protection Agency; ESH = Earth System History; FEMA = Federal Emergency Management Agency; ICDP = International Continental Drilling Program; LExEn = Life in Extreme Environments; MARGINS = Continental Margins Research; NASA = National Aeronautics and Space Administration; NEHRP = National Earthquake Hazards Reduction Program; NIST = National Institute of Standards and Technology; NOAA = National Oceanic and Atmospheric Administration; ODP = Ocean Drilling Program; OIA = Office of Integrative Activities (NSF); OPP = Office of Polar Programs (NSF); REU = Research Experience for Undergraduates; RPG = Research Planning Grants and career advancement; RUI = Research in Undergraduate Institutions; SCEC = Southern California Earthquake Center; USDA = U.S. Department of Agriculture; USGS = U.S. Geological Survey; WEAVE = Water and Energy: Atmospheric, Vegetative and Earth Interactions.

Figures A.1 and A.2 show the evolution of the EAR budget over the past 24 years. Since 1982, the number of programs within EAR has increased, a trend that accelerated in the 1990s. The budget for basic research has remained level but has been reprogrammed to accommodate two additional core programs (continental dynamics and hydrologic sciences) and 11 research-related special emphasis programs (see Table A.1). As a result, funding for the discipline core programs—the traditional heart of EAR—is at its lowest level in 20 years. Another significant change in the EAR budget occurred in the early 1990s, when funding priorities were shifted toward new facilities (e.g., Incorporated Research Institutions for Seismology [IRIS] and University NAVSTAR Consortium [UNAVCO]) and Science and Technology Centers (STCs). (These programs also have a significant research component.) The creation of these facilities and centers was in response to growing recognition in the 1980s of the creeping obsolescence of academic research facilities in the Earth sciences.[1-3] Most of these programs have been highly successful, and some have generated new money for EAR. For example, in 1997, the Department of Defense transferred funding for deployment of the Global Seismic Network (GSN) to EAR, and the Geosciences Division increased EAR funding for IRIS. More than 7.5 Terabytes of seismic data from the network and field experiments are now available through the IRIS data management system.

[1] *Opportunities for Research in the Geological Sciences*, National Academy Press, Washington, D.C., 95 pp., 1983.

[2] *Research Briefings 1983*, National Academy Press, Washington, D.C., 99 pp., 1983.

[3] *Earth Materials Research: Report of a Workshop on Physics and Chemistry of Earth Materials*, National Academy Press, Washington, D.C., 132 pp., 1987.

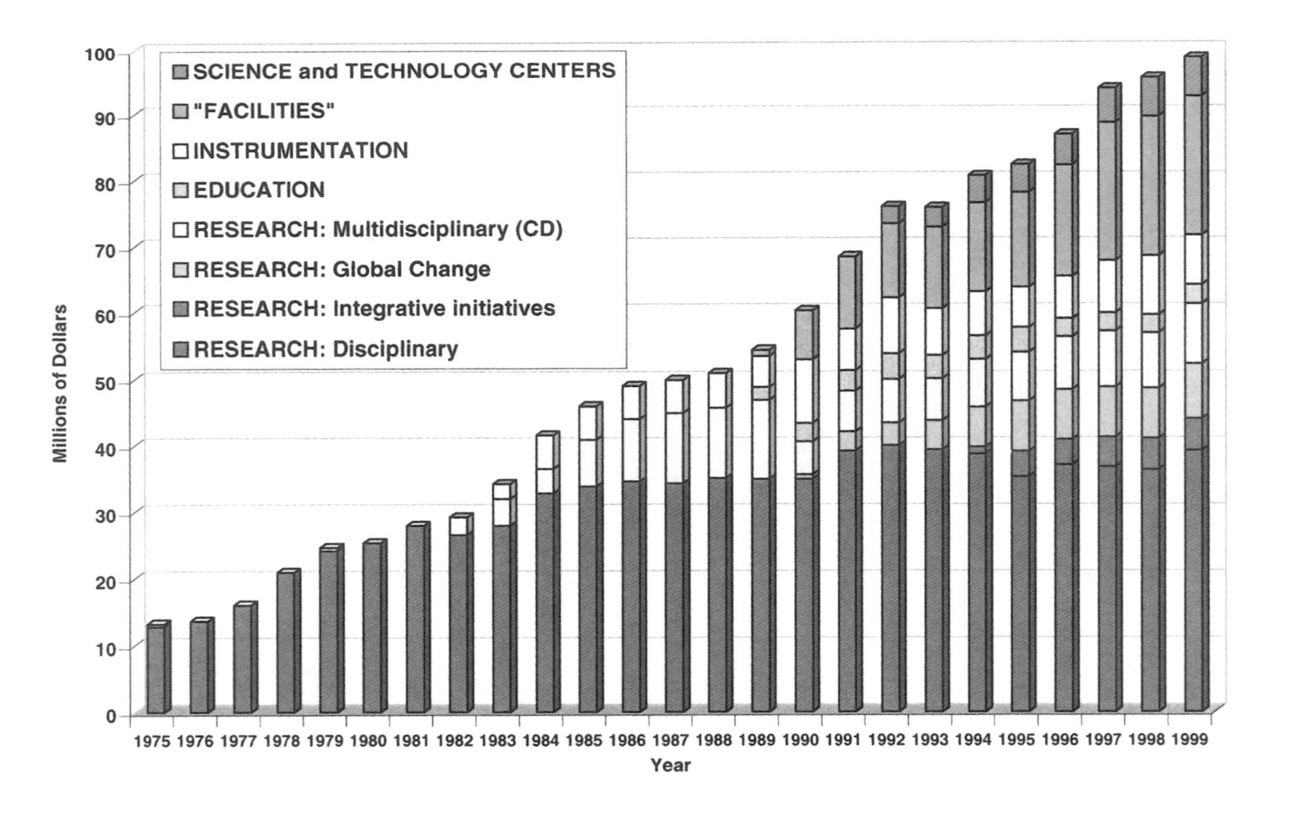

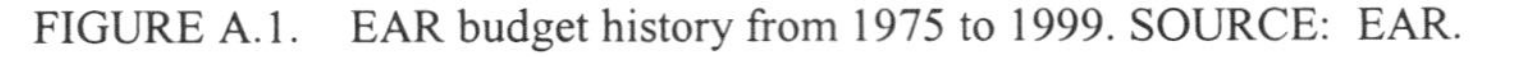

FIGURE A.1. EAR budget history from 1975 to 1999. SOURCE: EAR.

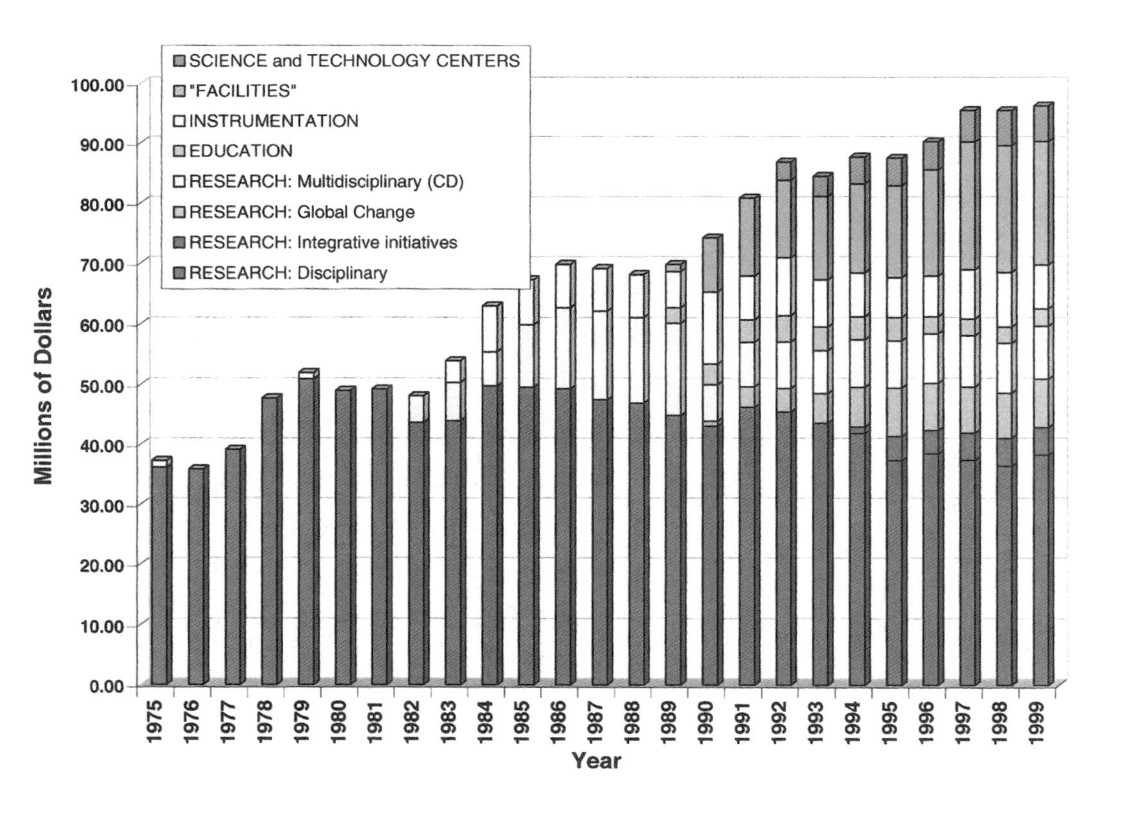

FIGURE A.2. EAR budget history from 1975 to 1999 in terms of constant 1998 buying power. SOURCE: EAR.

The FY 1999 budget of EAR is $95 million, with 62% devoted to research, 3% to education, and 35% to instruments, facilities, and science and technology centers. Although the budget has been increasing by an average of 4% per year (Figure A.1), the growth in the budget has not kept pace with the growth of the Earth science research community that relies on the NSF for funding. Between 1990 and 1998, the number of PIs applying for EAR funding rose by 17% (from 1949 to 2288), but the number of PIs funded fell by 10%. The success rate of a proposal to EAR is now 31%, down from 39% in 1990. Moreover, partly to support as many well-reviewed proposals as possible, EAR has not increased the size of its awards (on average $55,000 per award for research grants) for the last several years. Consequently, the buying power of EAR grants has decreased with time.

EAR Education Programs

EAR devotes most of its $1.8 million education budget to postdoctoral research fellowships and the NSF-wide Research Experience for Undergraduates (REU) Program. The latter both supplements individual investigator research grants and supports REU sites (see Table A.2), which bring students together in a common project. The REU program has been successful in getting students involved with cutting-edge research and pursuing careers in the Earth sciences. Many of the STCs host REU sites, and similar sites could be built into EarthScope or the natural laboratory program proposed in this report.

Two other NSF-wide programs in which EAR participates have been less successful—Faculty Early Career Development (CAREER) and Research in Undergraduate Institutions (RUI). The goal of the CAREER program is to reward young scientists who want to add a strong educational component to their research program. However, the proposal process places unrealistic expectations on junior faculty in terms of the required scope, and the proposals are judged against research proposals and therefore rarely succeed. The RUI program seeks to direct more research money to smaller schools which commonly have a greater emphasis on education than research universities. The proposals are given special consideration by the review panels and, if successful, will be funded from the relevant core program. Like other research proposals, RUI proposals are not required to have an explicit educational component and therefore do not further EAR goals in this area.

TABLE A.2. REU Sites Funded by EAR

Location	Task
University of Alaska	Field work in geology and geophysics in Alaska
University of Tennessee	Collaborations among students in social and physical sciences on environmental problems, science, and public policy
Georgia State University	Geologic and geochemical investigations through the Atlanta Consortium for Research in the Earth Sciences
University of Minnesota	Participation of women in research on the origin and history of glacial deposits
San Diego State University/Institute of Geophysics and Planetary Physics	Summer of Applied Geophysical Experience—Applied geophysics in the Rio Grande Rift
State University of New York, Stony Brook	Laboratory studies in high-pressure geophysics and mineral physics
Furman University	Interdisciplinary research on watersheds
Carleton College	Minority participation in the Keck Geology Consortium field program
University of Nevada, Reno	Geologic mapping in the field with mentors from the state geologic surveys
Carnegie Institution of Washington	Laboratory research in mineral physics and chemistry
National Museum of Natural History	Museum research training program

RELATED FEDERAL RESEARCH PROGRAMS

Several federal agencies support basic research in the Earth sciences. The research programs of the agencies with the largest Earth science programs—the U.S. Geological Survey (USGS), Department of Energy (DOE), and the National Aeronautics and Space Administration (NASA)—and their collaborations with EAR are described below and in Table A.3.

U.S. Geological Survey

The USGS is the nation's primary provider of Earth science information on geologic hazards and resources and the nation's geologic framework. The FY 1999 USGS budget was $1.1 billion, including appropriated funds and

TABLE A.3. Major Multiagency Collaborative Projects (FY 1999)

Project	EAR Contribution per Year	Partner Contribution per Year
APS synchrotron facility	Facility support ($1.4 million)	**DOE**: beamline construction, instrumentation support, Facility operations ($90 million), and beamline support ($0.7 million)
Continental drilling	Facility and research support ($1.2 million)	**USGS**: site characterization and research ($0.5 million)
Equipment[a]	Equipment purchase ($1.4 million)	**NASA**: equipment purchase ($2.4 million)
GSN	Network installation, operation and maintenance ($3.3 million)	**USGS**: network installation and service and data archive ($3.8 million)
NEHRP	Research and facilities ($10.9 million)	**USGS**: network installation, hazard assessment, and research ($50 million)
SAR data purchase	Data purchase and research support ($0.1 million)	**NASA**: data purchase and SAR consortium support ($0.1 million) **USGS**: data purchase ($0.05 million)
SCIGN array	Instrumentation and operations ($1.5 million)	**USGS**: instrumentation, personnel, and research support ($0.5 million) **NASA**: installation of 200 sites ($6 million), data processing ($0.1 million), and research support ($0.3 million)

UNAVCO	GPS equipment, engineering and technical support, data management system ($1.7 million)	**NASA**: research and support of GPS tracking stations ($1.2 million) and instrument development ($0.2 million)

[a] Includes ion microprobes, scanning electron microscopes, and thermal ionization and ion mass spectrometers.

NOTE: APS = Advanced Photon Source; SAR = synthetic aperture radar; SCIGN = Southern California Integrated GPS Array.

reimbursable contracts. Of this amount, $228 million was devoted to directed research, equipment and facilities in the Earth sciences. Major topics of research within the Geologic Division include geologic hazards, landscapes and coasts, and Earth resources:

- *Earthquake Hazards Program* supports research on earthquake characteristics, disseminates information to the public on the likelihood and potential effects of moderate to large earthquakes in high-risk regions, and assesses the risks of aftershocks and related ground motion following an earthquake. Major program elements include the National Earthquake Hazards Reduction Program (NEHRP) and the Global Seismic Network, a worldwide network of 107 seismographic stations.
- *Volcano Hazards Program* supports research on volcanic processes, assesses and monitors potential volcanic hazards, and provides warning information on volcanic activity. The program also supports four volcano observatories.
- *Landslide Hazards Program* supports research on landslide hazards in the urban environment, and landslides that occur in association with other natural disasters such as earthquakes. Applied research focuses on developing and deploying instrumentation to monitor potential landslides and forecast the onset of catastrophic movement.
- *Geomagnetism* supports research on understanding geomagnetic processes and their effects on our physical and social environment. The program also supports a network of 13 magnetic observatories, and a center to disseminate information on the Earth's changing magnetic field to the public.
- *Earth Surface Dynamics* supports research on surficial cycles and processes, rates of surface modifications, and the factors that control these rates of change, with an emphasis on natural and anthropogenic climate variability.
- *National Cooperative Geologic Mapping Program* supports geologic mapping studies and provides digital geologic maps to the public.
- *Coastal and Marine Geology Program* supports research on environmental quality and preservation of coastal and marine areas, and provides information on coastal erosion, storm effects, and offshore landslide, earthquake, and tsunami hazard potential to the public. The program also supports reconnaissance surveys and map production of the U.S. coast and Exclusive Economic Zone.
- *Mineral Resources Program* provides scientific information for resource assessments, including mineral potential, production, consumption, and impact of extraction and production on the environment.

• *Energy Resources Program* supports assessments of the quantity, quality, and geographic locations of natural gas, oil, and coal, and estimates of energy resource availability and recoverability.

Relevant programs within the Biological Resources Division include the National Biological Information Infrastructure as well as the following:

• *Biomonitoring of Environmental Status and Trends Program* supports research to identify and understand the effects of environmental contaminants on biological resources.
• *Land-Use History of North America Program* supports research on past and present changes in land cover and land use to better understand the planet's surface environment.

Key programs in the Water Resources Division include the following:

• *National Water-Quality Assessment* describes the status and trends in the quality of the nation's ground- and surface-water resources in order to provide a sound understanding of the natural and human factors that affect the quality of these resources.
• *Toxic Substances Hydrology Program* supports research on the contamination of surface water, groundwater, soil, sediment, and the atmosphere by toxic substances.
• *National Research Program* supports investigations on ground- and surface-water chemistry, ground- and surface-water hydrology, geomorphology, sediment transport, and ecology as they affect water resources.

EAR collaborations with USGS tend to focus on Earth science research, continental drilling, and arrays of geophysical instruments. Cooperation between the two agencies has been increasing in recent years, and the USGS sees future collaborative work on programs such as EarthScope and monitoring the physical properties of the Earth, including seismic, Global Positioning System (GPS), gravity, geomagnetic field networks, and observatories.

Department of Energy

The mission of DOE is to maintain the safety, security, and reliability of the U.S. nuclear weapons stockpile, without underground nuclear testing. DOE's FY 2000 budget is $17.8 billion, of which $165 million is devoted to basic Earth science research in geochemistry, geophysics, climate and hydrology, the carbon cycle, bioremediation, and surface remediation. An additional $2.2 billion is spent on applied programs that rely on the Earth sciences, including the Yucca Mountain Project, environmental management, fossil energy, geothermal energy development, and nonproliferation and verification. Major basic research programs include the following:

- *Geophysics and Earth Dynamics* supports research on large-scale Earth dynamics, evolution of geologic structures, properties of Earth materials, rock mechanics, fracture and fluid-flow, and underground imaging.
- *Geochemistry* supports research on thermochemical properties of geologic materials, rock-fluid interactions, organic geochemistry, and geochemical transport.
- *Energy Resource Recognition, Evaluation, and Utilization* supports research on resource definition and utilization, reservoir dynamics and modeling, properties and dynamics of magma, and continental scientific drilling.
- *Hydrogeology* supports research on fluid transport dynamics and modeling, thermochemical properties of energy materials, and perturbations of fluid flow.
- *Climate and Hydrology* supports research on global change.
- *Carbon Cycle* supports research related to understanding the geophysics and geochemistry of potential reservoirs appropriate for subsurface sequestration of CO_2 and identifying ways to enhance carbon sequestration in biomass and soils.
- *Bioremediation* supports research on the potential of microorganisms to remediate toxic chemicals at hazardous waste sites.
- *Subsurface Remediation* supports research related to transport modeling, contaminant geochemistry, and characterization of sites contaminated by nuclear weapons production and research.

These research activities are supported by a network of laboratories and national user facilities, such as synchrotron light sources, as well as by high-resolution microscopy, spectroscopy, and geophysical imaging tools.

EAR-DOE collaborations focus on synchrotron facilities, although DOE was also a major contributor to the continental drilling program until 1995. (Congress rescinded DOE funds because of a misperception that the program was being "double funded.") Although the increasing number of EAR programs is making it more difficult to identify joint projects, DOE anticipates future collaboration with EAR on DOE user facilities, such as an environmental molecular science beamline at the Advanced Light Source and new neutron science facilities such as the Spallation Neutron Source.

National Aeronautics and Space Administration

NASA conducts Earth science research related to the surface of the Earth, the Earth's interior, and other planets and solar system bodies. Research directed away from the Earth is conducted through the Office of Space Science, and research directed toward the Earth is conducted through the Office of Earth Science. The mission of the Space Science Enterprise is to solve mysteries of the universe, explore the solar system, discover planets around other stars, and search for life beyond Earth. Approximately $400 million (out of a total of $2.1 billion) was available in FY 1999 for peer-reviewed research and data analysis in these topics. Major Earth science research programs within the space sciences part of NASA include the following:

- *Planetary Geology and Geophysics Program* supports scientific investigations of planetary surfaces and interiors, satellites, and smaller solar system bodies such as asteroids and comets. The program also supports analysis of samples from planetary missions, such as lunar rocks and soils; geologic field studies of terrestrial analogues to planetary phenomena; lunar agriculture studies; and a wide array of analytical, experimental, modeling, and remote-sensing studies.
- *Cosmochemistry Program* supports cosmochemical investigations related to understanding the geochemical nature of solar system bodies and the formation and chemical development of the solar system. The program also supports equipment related to laboratory studies of extraterrestrial materials (meteorites, cosmic dust, and lunar samples), such as microprobes, scanning electron microscopes, and mass spectrometers.
- *Astrobiology Program* supports research on how life begins and develops, the possibility of life elsewhere in the universe, and the response of terrestrial life to environmental change and to conditions in space or on other planets.

The mission of the Earth Science Enterprise is to develop an understanding of the integrated Earth system (land, ocean, atmosphere, ice, and biota) and the effects of natural and human-induced changes on the global environment. The FY 1999 budget for the Earth Science Enterprise was $1.4 billion, of which $252 million was devoted to research and analysis. Major Earth science research programs include the following:

- *Laboratory for Terrestrial Physics* monitors plate motions and regions of intraplate deformation; develops improved gravity and topography models for the Earth, the moon, Mars, and Venus; measures time-variable gravity, postglacial rebound, and ice sheet thickness variations; and models the structure variations and sources of the Earth's main and crustal magnetic fields. The program also operates an international space geodetic network and provides precise celestial and terrestrial reference frames to the community.

- *Solid Earth and Natural Hazards Program* supports research related to geodynamics and geodesy, geopotential fields, geologic applications of remote sensing, and natural hazards. The program also supports the development and operation of global and regional geodetic networks and airborne data acquisition. The budget for this program was $24 million, of which half was devoted to research and half to infrastructure.
- *Land Surface Hydrology Program* supports research on the scientific basis of water resources management and the role of water in land-atmosphere interactions.
- *Global Energy and Water Cycle Experiment (GEWEX)* supports research to observe, understand, and model the hydrological cycle and energy fluxes in the atmosphere, at the land surface, and in the upper oceans. NASA also supports hydrologic studies in specific regions, such as the Amazon Basin and arid and semiarid lands.

EAR works with both the space science and the Earth sciences offices of NASA. Collaborative research tends to focus on technology development, equipment acquisition, and research related to large geophysical networks. Possible future collaborations identified by NASA program managers include the Plate Boundary Observatory, remote-sensing data purchases, and research in planetary volcanism. In their view, additional technology development is hampered by an NSF policy against funding soft-money researchers at federal centers such as the Jet Propulsion Laboratory.

Lessons Learned from Previous Collaborations with USGS, DOE, and NASA

Multiagency collaborations save money and permit broad scientific objectives to be reached that could not be achieved by an individual agency. Such collaborations may take the form of joint solicitations, commonly with joint panel evaluations, or each agency may fund investigators separately. The latter can be awkward to manage because the agencies are on different funding schedules and researchers may have to go through two peer review processes. NSF, USGS, DOE, and NASA program managers identified the following "lessons learned" in establishing joint projects:

- the best collaborations occur when there are a common interest, shared funding, and clear expectations of the results;
- funding and programmatic responsibilities of each agency should be delineated to reflect the culture of the agency (e.g., mission versus basic science, subject versus technique);
- good working relationships between program managers are essential; and
- agencies must be aware of perceptions in the community, such as which agency is getting the credit.

Potential Partnerships with Other Agencies and NSF Programs

The following federal agencies have programs with an Earth science component relevant to the research opportunities discussed in this report.

Department of Defense. The Department of Army, Office of Naval Research, Air Force Office of Scientific Research, and Advanced Applied Technology Demonstration Facility support research related to the Critical Zone, particularly detoxification of military bases and arsenals; refueling plants; land degradation; global soil surveys; and soil erosion, compaction, and trafficability (i.e., ability of soils to withstand a load).

Environmental Protection Agency (EPA). The EPA supports research related to geobiology, microorganisms in the environment, and the Critical Zone. In particular, the *Microbiological and Chemical Exposure Assessment Research Division* supports studies to determine the levels of hazardous chemical and microbials in environmental matrices and the environmental pathways by which hazardous contaminants are transported via air, water, food, and soil to populations at risk. EPA's *National Center for Environmental Research* sponsors research on environmental biology, environmental chemistry, and ecological effects of environmental stressors.

Federal Emergency Management Agency (FEMA). FEMA supports research related to the Critical Zone, particularly studies related to mitigation, response, recovery, and loss estimation for all natural hazards, as well as surveys and studies of flood hazard and erosion. FEMA also supports the multiagency *National Earthquake Hazard Reduction Program*.

National Institutes of Health (NIH). NIH supports research related to geobiology, the Critical Zone, and microorganisms in the environment. In particular, the National Institute of Environmental Health Sciences supports research on soil-borne pathogens, toxic chemicals, aerosol dusts, molecular biology, and the effect of soil consumption on human health.

National Oceanic and Atmospheric Administration (NOAA). NOAA supports research related to hydrology and the Critical Zone. In particular, the *Coastal Ocean Program* supports research aimed at characterizing changes in coastal areas (e.g., shorelines, habitat), reducing and reversing the degradation of coastal habitats, and predicting and assessing the impact of natural and human-induced hazards (including climate change) on coastal ecosystems and habitats. The *Office of Global Programs* supports the GEWEX Continental-Scale International Project, which examines the water-energy exchange processes involved in the coupling of the atmosphere and land surface in the Mississippi River Basin.

U.S. Department of Agriculture (USDA). The USDA supports research related to geobiology, microorganisms in the environment, and the Critical Zone. For example, the *Water Quality and Management Program* sponsors research on water and climate characterization, hydrologic processes, and watershed characteristics. The *Soil Resource Assessment and Management Program* supports research on soil conservation, carbon sequestration in soils, and soil properties and indicators. The *Global Change Program* supports research on hydrologic processes, agricultural greenhouse gas (carbon dioxide, methane, nitrous oxide) budgets, and the effect of climate change on the agriculture system. The *Natural Resources Conservation Program* supports the national soil survey mapping program and research on spatial diversity of soil-landscape patterns, wind, and water erosion processes; mechanisms of carbon sequestration in soil systems; soil quality; and wetland biogeochemical processes. The *Plant, Microbial, and Insect Genetic Resources, Genomics, and Genetic Improvement Program* supports research on microbial cycling of elements and associated new biotechnologies. Finally, the *Plant Biological and Molecular Processes Program* supports research on the responses of plant growth and development to the environment, and the nature of environmental, physical, and chemical messengers that trigger developmental changes in plants.

National Science Foundation. NSF supports research related to geobiology, Earth and planetary materials, the Critical Zone, EarthScope, and planetary science:

- *Atmospheric Sciences Division* supports research relevant to the Critical Zone, particularly the flux of trace gases into and out of the atmosphere, and the assembly and analysis of paleoclimate data.

• *Astronomical Sciences Division* supports research relevant to Earth and planetary materials and planetary science. In particular, the *Planetary Astronomy Program* supports theoretical and observational studies of the detailed structure and composition of planetary surfaces, interiors, atmospheres, and satellites; the nature of small bodies (asteroids and comets); and the origin and development of the solar system.

• *Biological Sciences Directorate* supports research relevant to geobiology and microorganisms in the environment. The *Division of Environmental Biology* supports research on the origins, functions, relationships, interactions, and evolutionary history of populations, species, communities, and ecosystems. The program also supports a network of long-term ecological research sites. The *Division of Integrative Biology and Neuroscience* supports research aimed at understanding the living organism (plant, animal, microbe) as a unit of biological organization. Such research encompasses (1) the integration of molecular, subcellular, cellular, and functional genomics approaches to understand the development, functioning, and behavior of organisms; and (2) the form and function of organisms in view of their evolution and environmental interactions. The *Division of Molecular and Cellular Biosciences* supports research related to understanding life processes at the molecular, subcellular, and cellular levels, including microbial biology and biomolecular materials. The *Microbial Observatories Initiative* focuses on the discovery and characterization of as-yet-undescribed microorganisms from diverse habitats. Research sponsored by the program includes properties and mechanisms responsible for microbial growth, adaptation, and survival in natural environments and microbial processes for anaerobic and aerobic flow of energy and cycling of nutrients, including aquatic, soil-rhizosphere, and sediment ecosystems. Finally, the infrastructure for research in biology—including instrumentation, research facilities, field stations, and computational biology—is supported by the *Division of Biological Infrastructure.*

• *Materials Research Division* supports theoretical and experimental research relevant to Earth and planetary materials, particularly in condensed-matter physics and nanomaterials, solid-state chemistry and polymers, and metals and ceramic materials.

• *Ocean Sciences Division* supports research and instrumentation development relevant to geobiology, the Critical Zone, and EarthScope. The *Biological Oceanography Program* supports research on the relationships between marine organisms and their interactions with the geochemical and physical environment, including molecular, cellular, and biochemical studies, evolutionary ecology, and sometimes systematic biology and paleoecology. The *Chemical Oceanography Program* supports research into the chemical components, reaction mechanisms, and geochemical pathways at the ocean-solid-Earth interface. The *Marine Geology and Geophysics (MG&G) Program* supports geology and geophysics research (e.g., structure, tectonics, volcanism, sedimentary processes, seawater or ocean rock geochemistry) on the ocean basins and margins, as well as the Great Lakes. It also covers interactions of continental and marine geologic processes. Important components of the MG&G program include (1) the Ridge Inter-Disciplinary Global Experiments, physical, chemical, and biological interactions between the midocean ridge volcanic system and the ocean environment, and (2) MARGINS (Continental Margins Research) (see Box A.2). Finally, the *Coastal Ocean Processes Program* supports research on the physical and meteorological processes that affect biological productivity, sedimentary processes, and chemical transformations in the coastal ocean system.

Appendix B

Community Input

SYMPOSIA

Geological Society of America Annual Meeting, 1998[1]

Ashley, G.M., Where are we headed? "Soft" rock research into the new millennium.
Baker, V.R., Adventitious and socially responsible research opportunities in the Earth sciences.
Bohlen, S.R., The Earth science century.
Cerling, T.E., and Ehleringer, J.R., Global biogeochemistry.
Gardner, W.R., Basic research opportunities in soil science from an Earth science perspective.
Hornberger, G.M., Linkages between the Earth and environmental sciences.
Kesler, S.E., Earth science research challenges: Water as a resource and a geologic agent.
Levander, A., Continental assembly, stability, and disassembly.
Simpson, D.W., Seismology—A cornerstone and keystone in Earth sciences.
Stanley, S.M., Paleontology and Earth system history in the new millennium.
Tauxe, L., and C. Constable, The role of paleomagnetism in the Earth sciences: Exciting new possibilities.

[1] *Research Opportunities in the Earth Sciences: A Ten-Year Vision*, Geological Society of America Annual Meeting, October 26-29, 1998.

American Geophysical Union Fall Meeting, 1998[2]

Ahrens, T.J., Shock wave, impact research and Earth science—Fortuitous application and research opportunities.

Boettcher, M., F.T. Wu, and E.A. Hetland, The Changbaishan, China PASSCAL Experiment: Perspectives from an IRIS intern.

Chave, A.D., J.R. Booker, and M.J. Unsworth, The magnetotelluric method in studies of continental tectonics.

Coe, R.S., Research opportunities in geomagnetism and paleomagnetism.

Crisp, D., and C.A. Raymond, NASA New Millennium Program: Space flight validation of advanced technologies for future science missions.

Fiske, R.S., Research opportunities in volcanology: A ten-year vision.

Freeman, K.H., Research opportunities in isotopic biogeochemistry: Methods, mechanisms and cross-disciplinary mysteries.

Geller, M.A., Atmosphere/solid Earth interface research.

Hager, B.H., Research opportunities and challenges in Earth sciences for the next decade.

Herring, T.A., Research opportunities in geodesy.

MacGregor, I., and S.W. Draheim, National Science Foundation: Support of research and education in the Earth sciences.

Metzger, E.P., Coordinating changes in Earth science research and science education for the benefit of both.

Orcutt, J.A., Global Earth science.

Park, J., From seismic properties to geologic processes: The next steps.

Potter, K.W., Hydrology and ecosystems.

Tanimoto, T., Surprise in Earth's oscillation.

Weller, R.A., The hydrological cycle—Linking the ocean and atmosphere to land.

Zimbelman, J.R., A planetary perspective on upcoming research opportunities in the Earth sciences.

---

[2]*Research Opportunities in the Solid-Earth Sciences: A 10-Year Vision*, American Geophysical Union Fall Meeting, December 6-10, 1998.

WORKSHOP REPORTS[3]

A Vision for Geomorphology and Quaternary Science Beyond 2000. Results of a workshop held in Minneapolis, Minnesota, February 6-7, 1999, 22 pp. (http://vishnu.glg.nau.edu/amqua/)

Participants: R. Anderson, A. Ashworth, T. Cerling; P. Clark, B. Dietrich, R. Graham, E. Grimm, V. Holliday, E. Ito, J. Knox, C. Oviatt, C. Paola, and L. Safran.

Coastal Sedimentary Geology Research a Critical National and Global Priority. Results of a workshop held in Honolulu, Hawaii, November 9-12, 1999, 11 pp.

Participants: J. Anderson, K. Crook, C. Fletcher, P. Larcombe, S. Rigss, A. Sallenger, D. Scott, I. Shennan, and R. Theiler.

Dynamic History of the Earth-Life System. Results of a workshop held in Washington, D.C., March 6-9, 1999, 19 pp. (http://tigger.cc.uic.edu:80/orgs/paleo/PaleoSocWorkshop.htm)

Participants: R. Bambach, P. Crane, S. D'Hondt, W. DiMichele, D. Erwin, K. Flessa, J. Flynn, R. Gastaldo, S. Holland, D. Jablonski, J. Jackson, R. Kaesler, P. Kelley, S. Kidwell, P. Koch, T. Lyons, C. Maples, C. Marshall, A. Miller, B. Gupta, D. Springer, and S. Stanley.

Geoscience Education: A Recommended Strategy. Results of a workshop held in Arlington, Virginia, August 29-30, 1996, 26 pp. (http://www.nsf.gov/pubs/1997/nsf97171/nsf97171.htm)

Participants: W. Bishop, L. Braile, S. Cook, L. Duguay, J. Hannah, R. Lopez, N. Marcus, M. Mayhew, J. Mitchell, D. Mogk, T. Moore, J. Prendeville, R. Ridky, R. Ryan, P. Samson, J. Snow, R. Somerville, D. Stephenson-Hawk, P. Stryker, Marilyn Suiter, and P. Wilkniss.

---

[3]The workshops listed were sponsored by the NSF or Earth science professional societies for a variety of purposes. The reports can be found on the committee's Web site *http://www4.nationalacademies.org/cger/besr.nsf* by clicking on "current studies," then on "Basic Research Opportunities in the Earth Science at the National Science Foundation." They may also be found at other Web addresses, as indicated in the reference given.

Microscopic to Macroscopic: Opportunities in Mineral and Rock Physics and Chemistry. Results of a workshop held in Scottsdale, Arizona, May 28-30, 1999, 18 pp.

Participants: T. Ahrens, J. Banfield, G. Brown, M. Brown, P. Dove, R. Ewing, H. Green, R. Hemley, R. Jeanloz, S. Karato, K. Leinenweber, H. Mao, G. Masters, A. Navrotsky, John Parise, C. Prewitt, L. Stixrude, D. Weidner, H. Wenk, and R. Wentzcovitch.

Opportunities in Low-Temperature and Environmental Geochemistry. Results of a workshop held in Boston, Massachusetts, June 5, 1999, 10 pp.

Participants: J. Banfield, J. Banner, C. Bethke, J. Blum, W. Casey, O. Chadwick, C. Eggleston, W. Elliott, K. Freeman, M. Goldhaber, H. Lane, T. Lyons, C. Mora, K. Nagy, D. Nordstrom, F. Phillips, C. Romanek, S. Savin, G. Sposito, A. Stone, M. Velbel, C. Yapp, and H. Zimmerman.

Research Priorities in the Geosciences: Philosophical Perspectives. Results of a workshop held in Boulder, Colorado, March 5-7, 1999, 5 pp.

Participants: K. Benammar, B. Foltz, R. Frodeman, P. Glazebrook, E. Hargrove, I. Klaver, A. Light, D. Michelfelder, C. Mitcham, M. Oelschlaeger, H. Rolston III, J. Rouse, I. Stefanovic, D. Strong, and P. Warshall.

Sedimentary Systems in Space and Time: High Priority NSF Research Initiatives in Sedimentary Geology. Results of a workshop held in Boulder, Colorado, on March 27-29, 1999, 7 pp.

Participants: G. Ashley, P. Flemings, P. Heller, D. Lowe, I. Montanez, D. Nummedal, C. Paola, T. Simo, R. Slingerland, D. Swift, and J. Syvitski.

Support for Research in Tectonics at NSF. White paper from the Division of Structural Geology and Tectonics, Geological Society of America, July 24, 1998, 5 pp. (http://www.geology.uiuc.edu/ SGTDiv/)

Participants: M. Brandon, D. Cowan, E. Moores, T. Pavlis, and J. Tullis.

LETTERS FROM INDIVIDUALS

Universities, Museums, and Professional Societies

Thomas H. Anderson, University of Pittsburgh
Jeff Bauer, Shawnee State University
Roger Bilham, University of Colorado
Grady Blount, Texas A&M University, Corpus Christi
Michael Brown, University of Maryland
Kevin Burke, University of Houston
Philip A. Candela, University of Maryland
Millard F. Coffin, University of Texas
Robert F. Diffendal, Jr., University of Nebraska, Lincoln
Ralph Franklin, Clemson University
M. Charles Gilbert, University of Oklahoma
D. Jay Grimes, University of Southern Mississippi
Stan Hart, Woods Hole Oceanographic Institution
Robert M. Hazen, Carnegie Institution of Washington
Attila Kilinc, University of Cincinnati
Mark Kuzila, University of Nebraska, Lincoln
Stephen Marshak, Geological Society of America, Division of Structure and Tectonics
Jim McWilliams, University of California, Los Angeles
J. William Miller, Jr., University of North Carolina, Asheville
Gary W. Petersen, Soil Science Society of America
James B. Phipps, Grays Harbor College
Richard D. Rosen, Atmospheric and Environmental Research, Inc.
Brian Schroth, San Francisco State University
Peter M. Sheehan, Milwaukee Public Museum
Andrew A. Sicree, Earth and Mineral Sciences Museum, Pennsylvania State University
Rudy Slingerland, Pennsylvania State University
Alvin J.M. Smucker, Michigan State University
Lee E. Sommers, Soil Science Society of America
Donald L. Sparks, Soil Science Society of America
Lee Suttner, Indiana University, Bloomington

Federal Agencies

D. James Baker, National Oceanic and Atmospheric Administration
Steven R. Bohlen, U.S. Geological Survey
Carson W. Culp, Bureau of Land Management
Charles G. Groat, U.S. Geological Survey
Carolita U. Kallaur, Minerals Management Service
P. Patrick Leahy, U.S. Geological Survey
Norine E. Noonan, Environmental Protection Agency
Horace Smith, Natural Resources Conservation Service
Randy Smith, National Imagery and Mapping Agency

Acronyms

AMS	accelerator mass spectrometry
AT	Active Tectonics (NSF)
ATM	Atmospheric Sciences Division (NSF)
BTU	British thermal unit
CAREER	Faculty Early Career Development (NSF)
CD	Continental Dynamics (NSF)
CMB	core-mantle boundary
CMS	Civil and Mechanical Systems Division (NSF)
CSEDI	Cooperative Studies of the Earth's Deep Interior (NSF)
CTBT	comprehensive test ban treaty
DOD	Department of Defense
DOE	Department of Energy
EAR	Earth Science Division (NSF)
EGB	Environmental Geochemistry and Biogeochemistry (NSF)
EHR	Education and Human Resources (NSF)
EPA	Environmental Protection Agency
ESH	Earth System History (NSF)
ESNL	Earth Science Natural Laboratory
FEMA	Federal Emergency Management Agency
GCM	global circulation model
GEWEX	Global Energy and Water Cycle Experiment
G&P	Geology and Paleontology (NSF)
GPS	Global Positioning System
GSN	Global Seismic Network
ICPMS	inductively coupled plasma mass spectrometry
I&F	Instrumentation and Facilities (NSF)

IGERT	Integrated Graduate Education and Research Training (NSF)
IMS	International Monitoring System
InSAR	interferometric synthetic aperture radar
IRIS	Incorporated Research Institutions for Seismology
IT^2	Information Technology for the Twenty-First Century (NSF)
KDI	Knowledge and Distributed Intelligence (NSF)
LExEn	Life in Extreme Environments (NSF)
LTER	Long Term Ecological Research (NSF)
MARGINS	Continental Margins Research (NSF)
MG&G	Marine Geology and Geophysics (NSF)
MGS	Mars Global Surveyor
MRE	Major Research Equipment (NSF)
NASA	National Aeronautics and Space Administration
NEHRP	National Earthquake Hazards Reduction Program
NIH	National Institutes of Health
NOAA	National Oceanic and Atmospheric Administration
NRC	National Research Council
NSF	National Science Foundation
OCE	Ocean Sciences Division (NSF)
PBO	Plate Boundary Observatory
PI	principal investigator
R&D	research and development
REU	Research Experience for Undergraduates (NSF)
RUI	Research in Undergraduate Institutions (NSF)
SAFOD	San Andreas Fault Observatory at Depth
UNAVCO	University NAVSTAR Consortium
USDA	U.S. Department of Agriculture
USGS	U.S. Geological Survey
WWSSN	World Wide Standardized Seismographic Network